# Fundamentals of Medical Ultrasonics

# Fundamentals of Medical Ultrasonics

Michiel Postema

*With contributions from*
Keith Attenborough
Michael J. Fagan
Odd Helge Gilja
Andrew Hurrell
Spiros Kotopoulis
Knut Matre
Annemieke van Wamel

*Foreword by*
Paul A. Campbell

**CRC Press**
Taylor & Francis Group
Boca Raton  London  New York

CRC Press is an imprint of the
Taylor & Francis Group, an **informa** business

A SPON PRESS BOOK

First published 2011 by Spon Press
2 Park Square, Milton Park, Abingdon, Oxfordshire OX14 4RN

Simultaneously published in the USA and Canada by Spon Press
711 Third Avenue, New York, NY 10017, USA

First issued in paperback 2017

*Spon Press is an imprint of the Taylor & Francis Group, an informa business*

Cover design: Stephen John Rees steverees64@mac.com

*British Library Cataloguing in Publication Data*
A catalogue record for this book is available from the British Library

*Library of Congress Cataloging in Publication Data*
Postema, Michiel, 1973-
Fundamentals of medical ultrasonics / Michiel Postema ; with contributions from
Keith Attenborough ... [et al.] ; foreword by Paul A. Campbell.
p. ; cm.
Includes bibliographical references.
1. Ultrasonics in medicine. 2. Diagnostic ultrasonic imaging. I. Attenborough, Keith.
II. Title. [DNLM: 1. Ultrasonics. 2. Biophysics. 3. Ultrasonic Therapy--methods. 4.
Ultrasonography--methods. QT 34]
R857.U48P67 2011
616.07'54--dc22
2010042782

ISBN13: 978-1-138-07723-2 (pbk)
ISBN13: 978-0-415-56353-6 (hbk)

# Foreword

Michiel Postema is one of life's more colourful characters. To the uninitiated,[1] this fact is perhaps most evident from his courageous sartorial displays during conference presentations. These fetching ensembles, however, represent much more than the whimsical eccentricities that sometimes pervade academic circles. Rather, my opinion is that the threads are simply an outward signature of an extreme inner confidence; confidence of the kind that comes about when one simply has a deeper, and broader, appreciation of the field than most of one's peer group. This is because Professor Postema is one of the most knowledgeable and capable young researchers in the ultrasound arena today.

'Ah, but you are his friend and are bound to say something like this, ..,' you may be thinking to yourself.

'Well, ... yes, ... and no,' is my reply. 'Yes, he is indeed my friend, ... but no, I am not bound to say anything other than the facts.'

So, I do not make the above statement lightly. I do it with the certainty, and indeed the gratitude, that comes from having personally tapped into Michiel Postema's knowledge and intuition on many [and several significant] occasions in the past. Furthermore, I state this in the knowledge that Postema's previous advice has gone some way towards guiding me towards excellence in my own career. Publication of this present volume now opens up the possibility that he might effect a similar service for the wider community! The most important question at this juncture, dear reader, is whether you will allow yourself the opportunity to be amongst the beneficiaries? Or to put it another way, is this book going to help you? Flip forward just beyond this foreword and consider the aspirational statement that Postema has selected from Peter Parker's comic strip alter-ego. Now, hold that thought — for we shall return to it at the end.

I first encountered Dr. Postema in 2005 at the Rotterdam Symposium on Ultrasound Contrast Agents. He had given an authoritative talk on contrast agent modelling, an area that my own group were trying to break into at the time, and later over drinks, we agreed to develop something of a collaboration in order to kick-start the programmes in our mutual areas. As is often the case, tokens of appreciation were exchanged upon return to our host institutions: in my opinion, I got the better deal. For the princely outlay of a bottle of (lesser known but nonetheless excellent) single malt which I despatched from Dundee, I received a copy of *Medical Bubbles*, the published version of Dr. Postema's PhD

---

[1] The initiated of course, that is, those who already know and love Michiel, will know how his 'colourful' nature can be manifested through many further possibilities — most of which are guaranteed to raise a smile.

thesis. The little book was a revelation, focussing as it did on the intricacies of high speed imaging of shelled microbubbles under well-controlled circumstances — exactly the area that my own group's experimental programme was hinged on. Within those hallowed pages, the practicalities had been addressed and explored exhaustively, including some specific, and as it transpired to us, hugely important nuances arising from the microscopy set-up. Moreover, analytical protocols were also addressed in detail. In short, the monograph was a veritable godsend to me and my PhD students. More than that, it was extremely well written in terms of brevity, relevance and clarity — the very model for a thesis — but more than that, the perfect foundation for a more complete and generalised work — one that would eventually evolve to become the present text.

So, what of the competition? The ultrasound/microbubble field has text books aplenty [*viz*, Recommended Reading, Appendix B] already, many of which are quite superb. Do we really need another? Yes, actually we do — and for the main reasons mentioned here below:

1. The pace of development for research in this area, especially in the *hot topic* areas, such as microbubble-based drug delivery, and advanced bubble-based diagnostics, is quite astounding. Any emerging monograph worth its salt simply must include judicious reference to the most recent papers so that foundational concepts are accurately contextualised, with appropriate caveats, in the most up to date fashion — *Fundamentals of Medical Ultrasonics* accomplishes this in some style.

2. Moreover, multi-disciplinary teams are increasingly favoured for (*e.g.* research council-funded) projects that fall at the interface of Physics/Engineering with Life Sciences/Medicine, and, as the author himself outlines in Chapter 1, with the perceived demand for technical and clinical ultrasound imaging on the rise, there is a clear need for a 'one-stop shop' that embraces all the relevant areas of expertise in an expansive fashion. *Fundamentals of Medical Ultrasonics* is the first book in recent times to include all these basic aspects of medical ultrasonics: including elasticity, vibrations, waves, acoustics, transducers, radiation, imaging, bubble physics, contrast agents and sonoporation. Moreover, as the biography section will testify [Appendix C], a diverse spectrum of experts has been assembled to inject the correct level of disciplinary detail in the most authoritative way. That said, the presence of Professor Postema as a co-author/editor on every chapter has also endowed the text with coherency and consistency throughout.

3. Finally, where the target audience embraces the advanced undergraduate, or postgraduate student population, *Fundamentals of Medical Ultrasonics* represents the first recent and completely up-to-date book with all equations derived completely and explicitly from first principles. This level of completeness will, it is hoped, find favour as a valuable student resource, appealing to both the mathematically rigourous, and also serving to walk

those who are less mathematically confident through the derivations in a formal way.

The book thus fills several niches that are not well served at the present time. Furthermore, the book is written in the most charming and idiosyncratic fashion that allows Professor Postema's character to shine though in each chapter. This is exemplified with that beginning quote from your friendly neighbourhood web-slinger, Mr Spider-Man. Is it appropriate I wondered, to have some comic-book super hero quotation in what otherwise is a serious scientific work? I questioned the author on what exactly he meant to evoke here, and had the reply that the 'power' in question was a metaphor for the output from ultrasound transducers, coupled with the 'responsibility' to achieve minimal collateral damage by finding the optimal operating parameters.

Perhaps, ..!

One thing *is* for sure, however, and that is that this initial quote simply starts the whole thought-provoking business within this monograph. An intriguing and uniquely considered page by page personal perspective on this compelling research area then ensues, one that will have you smiling as those pennies begin to drop and a new level of understanding begins to percolate through the little grey cells.

What more is there to say? If your desire is to get up to speed with modern ultrasonics in a medical context, together with the tools and techniques required to appreciate the nittiest grittiest of details, then I can wholeheartedly recommend this book. Use it. Enjoy it.

Dr. Paul Andrew Campbell
Reader in Physics & Royal Society Industry Fellow
University of Dundee
September 2010

*With great power comes great responsibility*

Spider-Man (2002)

# Contents

# 1

# Introduction

Ultrasonic imaging is an economic, reliable diagnostic technique. When taking into account absolute hospital operating expenses, X-ray and ultrasound have approximately the same price per examination. Other imaging techniques are roughly three times as expensive, except for catheterisation, which is twenty times as expensive. However, X-ray is a less desirable imaging technique than ultrasound, due to the negative ionising radiation effects. Therefore, novel ultrasound-based imaging techniques are being developed that may compete with other imaging techniques.[1]

*Fundamentals of Medical Ultrasonics* treats the physical and engineering principles of acoustics and ultrasound as used for medical applications. The book covers the basics of elasticity, linear acoustics, wave propagation, non-linear acoustics, transducer components, ultrasonic imaging modes, basics on cavitation and bubble physics, as well as the most common diagnostic and therapeutic applications. Its aim is to provide students and professionals in medical physics and engineering with a detailed overview of physical and technical aspects involved in medical ultrasonic imaging, whilst being a useful reference for clinical research staff.

Ultrasound has become the most used medical imaging modality in the German-speaking world and will no doubt become the most used modality world-wide. Hence, an increasing number of engineers working on medical ultrasonics development and implementation will need to be trained in this field. Not only does *Fundamentals of Medical Ultrasonics* satisfy this need, it also narrows the gap between the technician and the clinician who wants to know "how stuff really works".

---

[1]Postema M. Bubbles and ultrasound. *Appl Acoust* **2009** 70:1305.

## 1.1 Definition of sound

Sound waves are a form of mechanical vibration. They correspond to particle displacements in matter. Unlike electromagnetic waves, which can propagate in vacuum, sound waves need matter to support their propagation: a solid, a liquid, or a gas. The ear is an excellent acoustic detector in air but its sensitivity is limited to an interval between 20 Hz and 20 kHz. Audio-frequency sound is essential in communication and entertainment. The acoustics of buildings, particularly concert halls, has been the subject of considerable study. Unwanted audio-frequency sound is called noise. The study of noise and noise control is an important part of engineering.

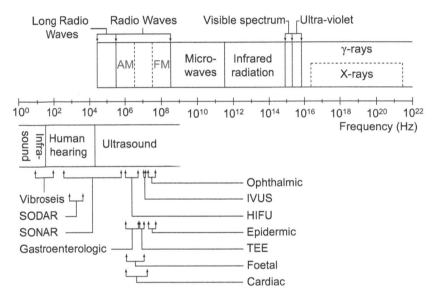

Figure 1.1: Typical applications of electromagnetic (*top*) and acoustic (*bottom*) frequencies.

Ultrasound refers to sounds and vibrations at frequencies above the upper audible limit of 20 kHz to values that can reach 1 GHz, as shown in Figure 1.1. Consequently, ultrasonics involve higher frequencies and smaller wavelengths than audio acoustics. The highest theoretical ultrasonic frequency that can be generated has been elegantly derived by Kuttruff.[2] Consider a crystal consisting of molecules separated by a distance $d$, through which a monotonous longitudinal wave with speed $c_p$ is travelling. The phase difference $\Delta\phi$ between two molecules is then

$$\Delta\phi = k\,d = 2\pi\,\frac{f\,d}{c_p}, \tag{1.1}$$

---

[2]Kuttruff H. *Acoustics: An introduction.* London: Taylor & Francis **2007**.

where $f$ is the sound frequency and $k$ is the wave number. If adjacent masses oscillate in opposite phase, the situation is that of a standing wave. Therefore, the highest theoretical ultrasonic frequency must be the frequency where $\Delta\phi = \pi$ or

$$f = \frac{c_\mathrm{p}}{2d}. \tag{1.2}$$

Conversely, infrasound involves sounds and vibrations at low frequencies (below 20 Hz) and long wavelengths. Because the physiological sensation of sound has disappeared at these frequencies, our perceptions of infrasound and ultrasound are different. Ultrasonic waves in fluids and solids are used for non-destructive evaluation. The general principle is to excite and detect a wave at ultrasonic frequencies and to deduce information from the signals detected. Example applications include the detection of flaws and inhomogeneities in solids, SONAR, SODAR, medical imaging, and acoustic microscopy.

For applications of acoustics, ultrasonics, and noise control, it is important to have a good understanding of the elasticity of materials and of wave propagation in infinite fluids and solids.

## 1.2   A brief history of cavitation and ultrasonics

One of the first to write on the concepts of cavitation was Leonardo da Vinci. Not only did he have a theory on the oversaturation of water:[3]

> *"Moreover the elements repel or attract each other, for one sees water expelling air from itself, and fire entering as heat under the bottom of a boiler and afterwards escaping in the bubbles on the surface of the boiling water."*

He was also the first to describe the concept of surface tension:

> *"The centre of a particular sphere of water is that which is formed in the tiniest particles of dew, which is often seen in perfect roundness upon the leaves of plants where it falls; it is of such lightness that it does not flatten out on the spot where it rests, and it is almost supported by the surrounding air, so that it does not itself exert any pressure, or form any foundation; and because of this its surface is drawn towards its centre with equal force and they become magnets one of another, with the result that each drop necessarily becomes perfectly spherical, forming its centre in the middle, equidistant from each point of its gravity, it always places itself in the middle between opposite parts of equal weight."*

Leonardo da Vinci had theories on acoustics, too:

---

[3]Richter IA, Ed. *The Notebooks of Leonardo da Vinci.* Oxford World's Classics paperback edition. New York: Oxford University Press **1998**.

*"I say that the sound of the echo is reflected to the ear after it has struck, just as the images of objects striking the mirrors are reflected into the eyes. And as the image falls from the object into the mirror to the eye at equal angles, so sound will also strike and rebound at equal angles as it passes from the first percussion in the hollow and travels out to meet the ear."*

As such, we may state that Leonardo da Vinci had already depicted the ingredients needed for contrast sonography five hundred years ago.

Not until the work of Daniel Bernoulli was it understood that negative pressure can be produced within a liquid.[4] Supposedly, Euler and d'Alembert debated the consequences of negative pressures in the early eighteenth century. Euler correctly assumed they would ultimately cause a rupture of the liquid, yet d'Alembert refused to accept this view.

Following the Industrial Revolution, as steam turbines became more powerful, the rotation speed of ship-screw propellers increased dramatically. With the increased rotation speed, an "extraordinarily rapid" kind of erosion of the propellers was observed.[5] Silberrad showed pictures of ship-screws and described the seriousness of this problem:

*"These have been photographed from specimens cut from the first set of propellers of the Mauretania, and exhibited at the recent Anglo-Japanese Exhibition. These original propellers had three loose blades, and an over-all diameter of about 17 ft. The metal, it will be seen, is eaten away to the depth of in some cases more than 2 in. The financial aspect of the question was thus very serious. Propellers of the alloys in question cost anywhere between 130l. to 180l. a ton and have a scrap value of less than half that amount, while the rapidity of the wear was such that in the case of the Mauretania and Lusitania they would, had no remedy been found, have required replacing every few months, at a cost of some thousands of pounds, since each propeller weighed about 20 tons."*

As for the cause of the erosion, Silberrad stated:

*"Further, it will be noted that the area attacked is, as has already been stated, near the hub. This was of large size, and it seems probable that there was a certain centrifugal action causing a reduction of pressure, and this region of reduced pressure was marked by the erosion. Here Dr. Silberrad considers that cavitation might occur, and, in consequence, water broken by intervening evacuated spaces with no air present."*

---

[4]Bernoulli D. *Hydrodynamica, sive de viribus et motibus fluidorum commentarii.* Strasbourg: JH Dulsecker **1738**.

[5]Silberrad O. Propeller erosion. *Engineering* **1912** 33–35.

The word *cavitation* for the formation of cavities due to negative pressures has been attributed to Froude.[6]

In 1917, Lord Rayleigh published his masterpiece on cavitation.[7] It starts with:

> *"When reading O. Reynolds's description of the sound emitted by water in a kettle as it comes to the boil, and their explanation as due to the partial or complete collapse of bubbles as they rise through cooler water, I proposed to myself a further consideration of the problem thus presented; but I had not gone far when I learned from Sir C. Parsons that he also was interested in the same question in connexion with cavitation behind screw-propellers, and that at his instigation Mr. S. Cook, on the basis of an investigation by Besant, had calculated the pressure developed when the collapse is suddenly arrested by the impact against a rigid concentric obstacle. During the collapse the fluid is regarded as incompressible.*
>
> *In the present note I have given a simpler derivation of Besant's results, and have extended the calculation to find the pressure in the interior of the fluid during the collapse. It appears that before the cavity is closed these pressures may rise very high in the fluid near the inner boundary."*

The equations presented are still applicable today.

Cavitation with a mechanical origin is called *hydraulic cavitation*. In a laboratory environment, de Haller experimented with hydraulic cavitation on turbines.[8] A more recent study on hydraulic cavitation involves snapping shrimp: predator shrimp that kill their prey by producing cavitation bubbles that collapse near their victims.[9]

Sound waves can create negative pressures, too, resulting into *acoustic cavitation*. To learn more about this subject, we have to look into the science of inaudible sound: *ultrasonics*.

In "Some Background History of Ultrasonics", Klein called ultrasonography, or: *superaudible acoustics*, "a by-product of the two world wars":[10]

> *"To trace the progress of ultrasonics from its beginning, it is necessary to recall the years 1914–1918 and Professor Paul Langevin who founded this subject. In 1915, the U-boat menace threatened the Allies. A Russian engineer named Chilowski submitted an idea for submarine detection to the French Government. The latter invited*

[6]Thornycroft JI, Barnaby SW. Torpedo-boat destroyers. *Min Proc Inst Civ Eng* **1895** 122:51–69.

[7]Lord Rayleigh. On the pressure development in a liquid during the collapse of a spherical cavity. *Philos Mag* **1917** 32:94–98.

[8]de Haller P. Untersuchungen über die durch Kavitation hergerufenen Korrosionen. *Schweiz Bauzeit* **1933** 101:243–246.

[9]Versluis M, Schmitz B, von der Heydt A, Lohse D. How snapping shrimp snap: through cavitating bubbles. *Science* **2000** 289:2114–2117.

[10]Klein E. Some background history of ultrasonics. *J Acoust Soc Am* **1948** 20:601–604.

*Langevin, then Director of the School of Physics and Chemistry in Paris, to evaluate it. Chilowski's proposal was to excite a cylindrical, mica condenser by a high frequency singing arc (Poulson Arc) operated at about 100 kc. This device was intended to be a generator of an ultrasonic beam for detecting submerged objects. The idea of locating underwater obstacles by means of sound echoes had been previously suggested by L.F. Richardson, in England. In 1912, following the Titanic disaster, he proposed to set a high frequency hydraulic whistle at the focus of a mirror and use the beam for locating submerged navigational hazards. Sir Charles Parsons, the inventor of the vapor turbine, became interested in this device and built one in accordance to Richardson's ideas. This apparatus was found unsuitable for the job of searching underwater obstacles, but the idea was not lost."*

(...)

*"By making use of the piezoelectric effects of quartz, Langevin introduced the modern piston transducer. He became acquainted with piezoelectric phenomena while a student at the laboratory of the Curie brothers. Perhaps the most outstanding advance made by Langevin in this field was his theoretical calculation and experimental verification of the fact that a thin sheet of quartz sandwiched between two steel plates constituted an electromechanical resonant system."*

(...)

*"Various aspects of the piezoelectric piston transducer were investigated by Langevin and his co-workers, among whom were a number of British and American scientists. They observed many biological and physical effects of the ultrasonic beam. For example, they noted in their laboratory tank that small fish were killed as they swam into the intense portion of the ultrasonic beam. Also, they saw incipient cavitation in the water when the sound source was active and felt painful effects upon the hand when struck in front of the beam."*

This is the first mentioning of acoustic cavitation.

But the cavities themselves produced sound waves as well. Bragg related the sound of drops falling into water to cavities:[11]

*"When photographs are taken from below the surface, it becomes clear that an air cavity is often formed."*

(...)

*"Now it appears that the note which we hear is the resonant note of this cavity, probably given out when the cavity has closed over*

[11]Bragg W. *The World of Sound*. London: G Bell and Sons **1920**.

*at the top and burst again. My friend Richard Paget has actually measured these cavities in various cases and then made models of them in plasticine. Blowing across the top of the model cavity, he finds that its note is practically the same as that the drop makes when it falls. The note is very high, and has a frequency of two to three thousand vibrations a second."*

In 1933, Minnaert presented his theory for the sounds created by bubbles in water:[12]

*"We will suppose that the bubbles give a sound because they pulsate in closing. Periodically the bubble expands and contracts, the surrounding water being the inert mass which is set in vibration, while the elasticity is due to the air of the bubble. A formula for the frequency of such pulsations may be derived in a quite elementary way."*

Thus, he formulated a theory on the resonance frequency of bubbles. Consequently, the resonance frequency of a free gas bubble in water is also referred to as the Minnaert frequency. Noltingk and Neppiras were the first to formulate an equation to describe the behaviour of gas-filled and empty cavities in a sound field.[13]

In 1954, shortly after the introduction of clinical ultrasound,[14,15] it was hypothesised that cavitation bubbles grow from gas nuclei encapsulated by organic skins. Because of the skins, these nuclei would not be subjected to diffusive processes.[16] This hypothesis could also be read as follows: If gas bubbles encapsulated by an elastic shell are sonicated, they may still act as cavitation nuclei.

Fourteen years later, experiments were done with saline to create air bubbles *in vivo.*[17,18] These rapidly diffusing air bubbles generated a characteristic response to ultrasound, such that perfused vessels would appear "brighter" on sonographic images. This new diagnostic technique was called ultrasound contrast imaging.

---

[12]Minnaert M. On musical air bubbles and the sound of running water. *Philos Mag* **1933** (S16):235–248.

[13]Noltingk BE, Neppiras EA. Cavitation produced by ultrasonics. *Proc Phys Soc London B* **1950** 63:674–685.

[14]Wild JJ. The use of ultrasonic pulses for the measurement of biologic tissues and the detection of tissue density changes. *Surgery* **1950** 27:183–188.

[15]Wild JJ, Neal D. Use of high-frequency ultrasonic waves for detecting changes of texture in living tissues. *Lancet* **1951** 655–657.

[16]Fox FE, Herzfield KF. Gas bubbles with organic skins as cavitation nuclei. *J Acoust Soc Am* **1954** 26:984–989.

[17]Gramiak R, Shah PM. Echocardiography of the aortic root. *Invest Radiol* **1968** 3:356–366.

[18]Gramiak R, Shah PM, Kramer DH. Ultrasound cardiography: contrast studies in anatomy and function. *Radiology* **1969** 92:939–948.

## 1.3   Outline and acknowledgements

The historic background in Chapter 1 has been based on the chapter "On Cavitation, High-speed Photography, and Medical Bubbles" in *Medical Bubbles*.[19]

Chapter 2 treats the basics of stress and strain analysis, including stress tensors, principal stresses, strain, Hooke's law, and stress functions. It has been based on the Stress and Strain Analysis lectures by Michael J. Fagan and Michiel Postema at The University of Hull.[20]

Chapter 3 covers the basics of vibrations, including mass–spring–dashpot systems. It has been based on the Dynamics lectures by Keith Attenborough and Michiel Postema at The University of Hull.[21]

Chapter 4 continues with waves, sound propagation, reflection, and transmission. It has been based on the Acoustics lectures by Keith Attenborough and Michiel Postema at The University of Hull.[22]

Chapter 5 describes the physics and engineering of ultrasound transducers. It has been written by Andrew Hurrell.

Chapter 6 treats the theory of radiated fields. It has been written by Andrew Hurrell.

Chapter 7 gives an overview of clinical ultrasonic imaging techniques. It has been written by Knut Matre and Odd Helge Gilja.

Chapter 8 thoroughly analyses bubble dynamics relevant for the use of ultrasound contrast agents. We kindly acknowledge the American Association of Physicists in Medicine for permission to reprint Figures 8.5 and 8.11, Bentham Science Publishers for permission to reprint Figure 8.3, and Elsevier Science for permission to reprint excerpts from several scientific papers and Figures 8.6 and 8.7.

Chapter 9 describes some of the imaging and therapeutic applications of contrast-enhanced ultrasonics. We kindly acknowledge Bentham Science for permission to reprint excerpts from a scientific paper and Expert Reviews for permission to reprint Table 9.1.

When designing the cover, Stephen John Rees was inspired by equation 3.12. Throughout this book, the illustrations were drawn, adjusted or pimped-up by Spiros Kotopoulis.

[19]Postema MAB. *Medical Bubbles*. S.l.: s.n. **2004**.

[20]Fagan MJ, Postema. *An introduction to stress and strain analysis*. Bergschenhoek: Postema **2007**.

[21]Attenborough K, Postema M. *A pocket-sized introduction to dynamics*. Bergschenhoek: Postema **2008**.

[22]Attenborough K, Postema M. *A pocket-sized introduction to acoustics*. Bergschenhoek: Postema **2008**.

# 2

# Stress, strain, and elasticity
## with Michael J. Fagan

The mathematical formulation of wave propagation in solids involves the use of the concepts and principles of 3-dimensional stress and strain analysis. Hence, we start by outlining these principles.

## 2.1 The uniform state of stress

Consider a continuous 3-dimensional body subjected to an arbitrary system of forces. The state of stress at a point O in the body can be studied by considering an infinitesimal parallelepiped erected at this point. It is assumed that the resultant forces acting on any two parallel faces are the same, *i.e.*, a uniform state or field of stress exists.

The isolated element is shown in Figure 2.1 referred to a Cartesian coordinate system. The double subscript is interpreted as follows: The first subscript indicates the direction of a normal to the plane or face on which the stress component acts; the second subscript relates to the direction of the stress itself. Note that $\sigma_x \equiv \sigma_{xx}$. Thus, $\tau_{xy}$ is the shear stress on the $x$-face in the $y$-direction.

The following sign convention is used: If face (F) and direction (D) are both positive, $\tau_{FD}$ is positive; if F and D are both negative, $\tau_{FD}$ is positive; if F and D are of opposite signs, $\tau_{FD}$ is negative.

By taking moments of the forces due to the stresses about each axis, we can show that

$$\begin{aligned}
\tau_{xy} &= \tau_{yx} \; ; \\
\tau_{xz} &= \tau_{zx} \; ; \\
\tau_{yz} &= \tau_{zy} \; .
\end{aligned} \tag{2.1}$$

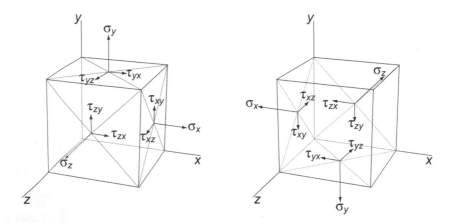

Figure 2.1: Stresses acting on the positive *(left)* and negative *(right)* faces of an infinitesimal body.

In future therefore, no distinction will be made between $\tau_{xy}$ and $\tau_{yx}$, $\tau_{xz}$ and $\tau_{zx}$, or $\tau_{yz}$ and $\tau_{zy}$. This means that only 6 Cartesian components are necessary for the complete specification of the state of stress at any point in the body. These terms can be conveniently assembled into the so-called stress tensor:

$$[\sigma] = \begin{bmatrix} \sigma_x & \tau_{xy} & \tau_{xz} \\ \tau_{yx} & \sigma_y & \tau_{yz} \\ \tau_{zx} & \tau_{zy} & \sigma_z \end{bmatrix}. \tag{2.2}$$

## 2.2   Stress on an inclined plane

It is required to find the state of stress on a plane inclined to the axes previously setup, represented by the face ABC of the tetrahedron in Figure 2.2, assuming that the stresses on faces OBC, OCA, and OAB are known. The position of the plane can be specified by the length and orientation of the normal OD drawn from the origin O to the plane ABC such that the angles ODA, ODB, and ODC are all right angles. The length of OD is equal to $r$, and its position is given by the angles AOD, BOD, and COD. The cosines of these angles are known as direction cosines and are denoted by

$$\begin{aligned} \cos \text{ AOD} &= l \ ; \\ \cos \text{ BOD} &= m \ ; \\ \cos \text{ COD} &= n \ . \end{aligned} \tag{2.3}$$

It can be proven on geometrical grounds, that

$$l^2 + m^2 + n^2 = 1, \tag{2.4}$$

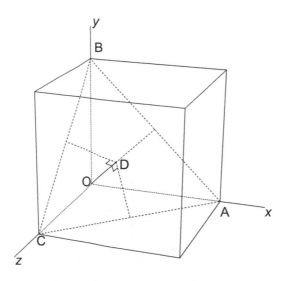

Figure 2.2: Plane inclined to the axes, represented by the face ABC of a tetrahedron.

so that only two of the direction cosines are independent.

The areas of the perpendicular planes OBC, OCA, and OAB may now be found in terms of these direction cosines:

$$OA = \frac{r}{l} ;$$

$$OB = \frac{r}{m} ;$$ 

$$OC = \frac{r}{n} .$$

(2.5)

Let the area of face ABC be $A$, and that of OCA be $A_y$. The volume of the tetrahedron can be written as $\frac{1}{3}Ar = \frac{1}{3}A_y OB = \frac{1}{3}A_y \frac{r}{m}$, from which it follows that $A_y = A\,m$. Hence,

$$\frac{A_x}{l} = \frac{A_y}{m} = \frac{A_z}{n} = A.$$

(2.6)

These are the relationships between the areas of the four faces of the tetrahedron.

With the stress on the inclined plane represented by its three Cartesian components $s_x$, $s_y$, and $s_z$, the general state of stress on the tetrahedron is shown in Figure 2.3. The equilibrium of the element can be examined by resolving the forces acting on it in the directions of the three axes. For example, in the $x$-direction,

$$s_x A - \sigma_x A_x - \tau_{yx} A_y - \tau_{zx} A_z = 0.$$

(2.7)

Hence,

$$s_x = \sigma_x l + \tau_{yx} m + \tau_{zx} n.$$

(2.8)

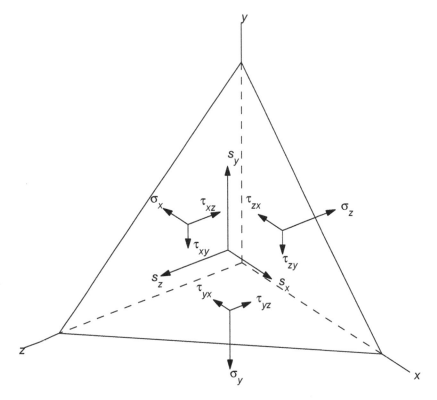

Figure 2.3: State of stress on a tetrahedron.

Similar expressions exist for $s_y$ and $s_z$. These three equations can be expressed in matrix form:

$$
\begin{bmatrix} s_x \\ s_y \\ s_z \end{bmatrix} = \begin{bmatrix} \sigma_x & \tau_{xy} & \tau_{xz} \\ \tau_{yx} & \sigma_y & \tau_{yz} \\ \tau_{zx} & \tau_{zy} & \sigma_z \end{bmatrix} \begin{bmatrix} l \\ m \\ n \end{bmatrix}. \tag{2.9}
$$

The resultant stress on the inclined plane is given by the resultant forces acting on ABC divided by the area of ABC. Therefore,

$$
s = \sqrt{s_x^2 + s_y^2 + s_z^2}. \tag{2.10}
$$

To find the normal stress on the plane, the forces parallel to the normal have to be resolved, noting that the area ABC is common to all forces acting on this face:

$$N = s_x l + s_y m + s_z n = \begin{bmatrix} l & m & n \end{bmatrix} \begin{bmatrix} s_x \\ s_y \\ s_z \end{bmatrix} =$$

$$= \begin{bmatrix} l & m & n \end{bmatrix} \begin{bmatrix} \sigma_x & \tau_{xy} & \tau_{xz} \\ \tau_{yx} & \sigma_y & \tau_{yz} \\ \tau_{zx} & \tau_{zy} & \sigma_z \end{bmatrix} \begin{bmatrix} l \\ m \\ n \end{bmatrix}.$$

$$(2.11)$$

The square of the resultant stress is equal to the sum of the squares of the normal stress and the shear stress on the plane, so that

$$T = \sqrt{s^2 - N^2}. \qquad (2.12)$$

The direction cosines of this resultant shear stress may also be found. Let $l_1$, $m_1$, and $n_1$ be the direction cosines of $T$ with respect to XYZ. Then, in the $x$-direction,

$$s_x A = N\,A\,l + T\,A\,l_1. \qquad (2.13)$$

Therefore,

$$l_1 = \frac{s_x - N\,l}{T} ;$$

$$m_1 = \frac{s_y - N\,m}{T} ; \qquad (2.14)$$

$$n_1 = \frac{s_z - N\,n}{T} .$$

## 2.3 Transformation of stresses for rotation of axes

By extending the theory developed in Section 2.2, it is possible to transform the tensor of a given state of stress known for one set of axes to the tensor of the same state in a second set of axes. For example, if the stress tensor in the $(x, y, z)$ coordinate system is $[\sigma]$, it is possible to determine the stress tensor $[\sigma']$ in another coordinate system $(x', y', z')$.

The relative rotation of the second system to the first is defined by 9 direction cosines, where for example $l_1$ is the angle between the $x'$-axis and the $x$-axis, $m_1$ is the angle between the $x'$-axis and the $y$-axis, and $l_2$ is the angle between the $y'$-axis and the $x$-axis.

The full set of cosines is

$$[L] = \begin{bmatrix} l_1 & m_1 & n_1 \\ l_2 & m_2 & n_2 \\ l_3 & m_3 & n_3 \end{bmatrix}. \qquad (2.15)$$

Consider a second Cartesian system $(x', y', z')$ and the set used in Section 2.2. Let the $x'$-axis be coincident with the normal of the plane that was examined: $\sigma_{x'} = N$. The normal stress on the plane follows from (2.11):

$$\sigma_{x'} = N = \begin{bmatrix} l_1 & m_1 & n_1 \end{bmatrix} \begin{bmatrix} \sigma_x & \tau_{xy} & \tau_{xz} \\ \tau_{yx} & \sigma_y & \tau_{yz} \\ \tau_{zx} & \tau_{zy} & \sigma_z \end{bmatrix} \begin{bmatrix} l_1 \\ m_1 \\ n_1 \end{bmatrix}. \tag{2.16}$$

By using the same method for the other axes, the other components of the stress tensor $[\sigma']$ can be determined:

$$[\sigma'] = [L][\sigma][L]^{\mathrm{T}}. \tag{2.17}$$

Using this equation, a tensorial state of stress in one coordinate system can be easily converted into another system.

## 2.4 Principal stresses

A principal plane is defined as one on which the shear stress is zero. The normal stress on this plane is known as the principal stress and is denoted by $p$. If the direction cosines of the plane are $l$, $m$, and $n$, then resolving stresses on the plane in the coordinate directions gives

$$\begin{bmatrix} s_x \\ s_y \\ s_z \end{bmatrix} = \begin{bmatrix} p\,l \\ p\,m \\ p\,n \end{bmatrix} = \begin{bmatrix} p & 0 & 0 \\ 0 & p & 0 \\ 0 & 0 & p \end{bmatrix} \begin{bmatrix} l \\ m \\ n \end{bmatrix}. \tag{2.18}$$

Rearranging this with (2.9) gives

$$\begin{bmatrix} \sigma_x - p & \tau_{xy} & \tau_{xz} \\ \tau_{yx} & \sigma_y - p & \tau_{yz} \\ \tau_{zx} & \tau_{zy} & \sigma_z - p \end{bmatrix} \begin{bmatrix} l \\ m \\ n \end{bmatrix} = 0. \tag{2.19}$$

This matrix represents three linear equations in $l$, $m$, and $n$, which will have non-trivial solutions if and only if

$$\det \|[\sigma] - [p]\| = 0. \tag{2.20}$$

Or, after expansion of the determinant

$$\begin{aligned} p^3 &- (\sigma_x + \sigma_y + \sigma_z)p^2 \\ &+ (\sigma_x\sigma_y + \sigma_y\sigma_z + \sigma_z\sigma_x - \tau_{xy}^2 - \tau_{yz}^2 - \tau_{zx}^2)p \\ &- (\sigma_x\sigma_y\sigma_z + 2\tau_{xy}\tau_{yz}\tau_{xz} - \sigma_x\tau_{yz}^2 - \sigma_y\tau_{zx}^2 - \sigma_z\tau_{xy}^2) = 0. \end{aligned} \tag{2.21}$$

This is known as the stress cubic. Its roots, which are always real, are the values of the three principal stresses, $p_1$, $p_2$, and $p_3$, which exist on three perpendicular planes. Note that the equation is independent of the direction cosines and therefore the coordinate system. If a different set of axes had been used to describe the element at point O, different values of applied stress $\sigma_x$, $\tau_{xy}$, ... would have resulted. However, the principal stresses would remain unaltered and the same stress cubic would have been derived. This means that the coefficients are the same, regardless of the original choice of axes, $i.e.$, they are invariant. By convention, the coefficients of the stress cubic are referred to as first, second, and third stress invariants:

$$
\begin{aligned}
I_1 &= \sigma_x + \sigma_y + \sigma_z \ ; \\
I_2 &= \sigma_x\sigma_y + \sigma_y\sigma_z + \sigma_z\sigma_x - \tau_{xy}^2 - \tau_{yz}^2 - \tau_{zx}^2 \ ; \\
I_3 &= \sigma_x\sigma_y\sigma_z + 2\tau_{xy}\tau_{yz}\tau_{xz} - \sigma_x\tau_{yz}^2 - \sigma_y\tau_{zx}^2 - \sigma_z\tau_{xy}^2 \ .
\end{aligned}
\tag{2.22}
$$

Thus, the stress cubic can be rewritten in terms of $p$ and I:

$$
p^3 - I_1 p^2 + I_2 p - I_3 = 0.
\tag{2.23}
$$

and, as the stress invariants have the same values regardless of the axes chosen, they must also be the same when the axes of reference correspond to the axes perpendicular to the principal planes. Therefore,

$$
\begin{aligned}
I_1 &= p_1 + p_2 + p_3 \ ; \\
I_2 &= p_1 p_2 + p_2 p_3 + p_3 p_1 \ ; \\
I_3 &= p_1\, p_2\, p_3 \ .
\end{aligned}
\tag{2.24}
$$

By convention, it is assumed that $p_1 > p_2 > p_3$.

One further invariant dubbed $I_2'$ is derived from $I_1$ and $I_2$:

$$
I_2' = 2(I_1^2 - 3I_2).
\tag{2.25}
$$

From (2.22),

$$
I_2' = (\sigma_x - \sigma_y)^2 + (\sigma_y - \sigma_z)^2 + (\sigma_z - \sigma_x)^2 + 6(\tau_{xy}^2 + \tau_{xz}^2 + \tau_{yz}^2),
\tag{2.26}
$$

or, in terms of principal stresses,

$$
I_2' = (p_1 - p_2)^2 + (p_2 - p_3)^3 + (p_3 - p_1)^2.
\tag{2.27}
$$

The so-called Von Mises yield stress $\sigma_{ys}$ is related to $I_2'$ as follows:

$$
\sigma_{ys}^2 = \frac{I_2'}{2}.
\tag{2.28}
$$

$I_1$ and $I_2'$ are of particular interest to us because they are related to the octahedral normal and octahedral shear stress, respectively.

The directions of the principal stresses can be found by substituting each of the three values of $p$ in turn into (2.19) and, using (2.4), solving for each set of $l$, $m$, and $n$.

## 2.5  Stationary values of shear stress

Shear planes contain maximum shear stresses corresponding to two of the principal stresses (*cf.* the 2-dimensional situation, where $T_{\max} = \frac{1}{2}(p_1 - p_2)$, evident from Mohr's circle of stress in Figure 2.4). Therefore, there will be three of these planes and their corresponding stresses.

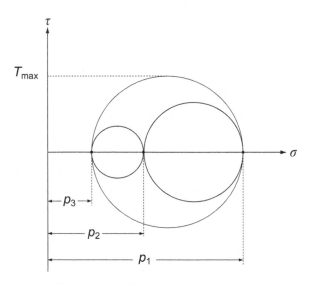

Figure 2.4: Mohr's circle of stress.

Having determined the positions of the principal planes and values of the principal stresses, then rotate the Cartesian coordinate system to correspond to the directions of the normals of the principal planes, as demonstrated in Figure 2.5. The stress components on the plane are

$$\begin{bmatrix} s_x \\ s_y \\ s_z \end{bmatrix} = \begin{bmatrix} p_1 & 0 & 0 \\ 0 & p_2 & 0 \\ 0 & 0 & p_3 \end{bmatrix} \begin{bmatrix} l \\ m \\ n \end{bmatrix} = \begin{bmatrix} p_1\, l \\ p_2\, m \\ p_3\, n \end{bmatrix}. \tag{2.29}$$

The resultant stress is

$$s = \sqrt{p_1^2 l^2 + p_2^2 m^2 + p_3^2 n^2}, \tag{2.30}$$

and the normal stress is

$$N = p_1 l^2 + p_2 m^2 + p_3 n^2. \tag{2.31}$$

Therefore, the shear stress is given by

$$T^2 = S^2 - N^2 = (p_1^2 l^2 + p_2^2 m^2 + p_3^2 n^2) - (p_1 l^2 + p_2 m^2 + p_3 n^2)^2. \tag{2.32}$$

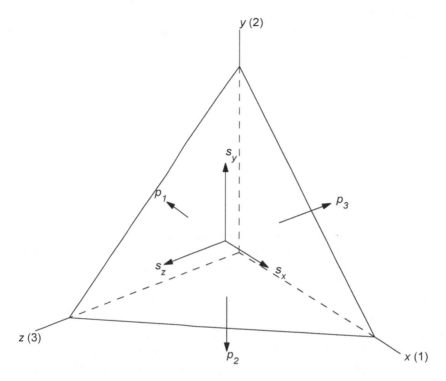

Figure 2.5: State of stress after rotation of the Cartesian coordinate system.

Using (2.4), one direction cosine can be eliminated:

$$\begin{aligned} T^2 = \quad & l^2(p_1^2 - p_3^2) + m^2(p_2^2 - p_3^2) + p_3^2 \\ & - \left[l^2(p_1 - p_3) + m^2(p_2 - p_3) + p_3\right]^2 . \end{aligned} \tag{2.33}$$

The values of $l$ and $m$, and therefore $n$, that maximise or minimise $T$ are found by differentiating (2.33) with respect to $l$ and $m$ and equating to zero:

$$T\frac{\partial T}{\partial l} = l(p_1 - p_3)\left\{(p_1 + p_3) - 2\left[l^2(p_1 - p_3) + m^2(p_2 - p_3) + p_3\right]\right\} = 0;$$

$$T\frac{\partial T}{\partial m} = m(p_2 - p_3)\left\{(p_2 + p_3) - 2\left[l^2(p_1 - p_3) + m^2(p_2 - p_3) + p_3\right]\right\} = 0. \tag{2.34}$$

Clearly, these equations vanish if $l = m = 0$ and $n = 1$, but this locates the plane on which $p_3$ acts, which by definition is a principal plane on which the shear stress $T = 0$. The other minima can similarly be found on the other principal planes. The maxima are found as follows. Assume $l = 0$. This satisfies the first equation in (2.34). The second will vanish if

$$(p_2 + p_3) - 2\left[m^2(p_2 - p_3) + p_3\right] = 0, \tag{2.35}$$

or, simplified, if

$$(p_2 - p_3)(1 - 2m^2) = 0. \tag{2.36}$$

This holds for

$$m = \pm \frac{1}{\sqrt{2}}. \tag{2.37}$$

Thus,

$$n = \pm \frac{1}{\sqrt{2}}. \tag{2.38}$$

Substituting back into (2.32), the shear is found to be

$$T = \frac{1}{2}(p_2 - p_3). \tag{2.39}$$

By a similar approach, the other shear planes and shear stresses can be found; the results are summarised in Table 2.1. The planes on which the maximum shear stresses act are illustrated in Figure 2.6. Note for instance that the shear stress $\frac{1}{2}(p_1 - p_2)$ acts on a plane given by $l = \pm\frac{1}{\sqrt{2}}$, $m = \pm\frac{1}{\sqrt{2}}$, $n = 0$. The normal to this plane is at right angles to the $z$ (3) axis and bisects the right angle between the $x$ (1) and $y$ (2) axes.

Corresponding to each shear stress, there is a normal stress (2.31). With $l = m = \frac{1}{\sqrt{2}}$ and $n = 0$, $N = \frac{1}{2}(p_1 + p_2)$. Therefore, on this principal plane of shear, there is a shear stress $T = \frac{1}{2}(p_1 - p_2)$ and a normal stress $N = \frac{1}{2}(p_1 + p_2)$.

## 2.6   Octahedral stresses

It is possible to find a more meaningful indication of the state of stress, rather than the tensorial description, by considering the octahedral sections of the element. These are planes equally inclined to the principle stress axes. From consideration of Figure 2.7 it is obvious that each of the eight planes will be subjected to the same value of direct stress and the same value of shear stress. This means that whereas it took six parameters to describe the state of stress in a set of rectangular sections, it takes only two to describe the magnitudes (though not directions) of the stresses on octahedral sections. Note that octahedral

|  | $T_{\min}$ | | | $T_{\max}$ | | |
|---|---|---|---|---|---|---|
| $l$ | 0 | 0 | $\pm 1$ | 0 | $\pm\frac{1}{\sqrt{2}}$ | $\pm\frac{1}{\sqrt{2}}$ |
| $m$ | 0 | $\pm 1$ | 0 | $\pm\frac{1}{\sqrt{2}}$ | 0 | $\pm\frac{1}{\sqrt{2}}$ |
| $n$ | $\pm 1$ | 0 | 0 | $\pm\frac{1}{\sqrt{2}}$ | $\pm\frac{1}{\sqrt{2}}$ | 0 |
| $T$ | 0 | 0 | 0 | $\frac{1}{2}(p_2 - p_3)$ | $\frac{1}{2}(p_1 - p_3)$ | $\frac{1}{2}(p_1 - p_2)$ |

Table 2.1: Minimum and maximum shear stresses and corresponding direction cosines.

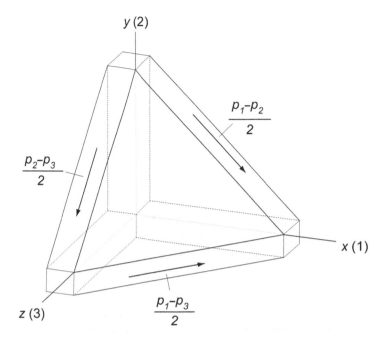

Figure 2.6: Planes of maximum shear stresses.

normal and shear stresses correspond to two fundamental effects of uniform dilation and uniform shear.

The magnitudes of the octahedral stresses are easily obtained. If the direction cosines of the normal to the octahedral plane are $l$, $m$, and $n$, their values must be $l^2 = m^2 = n^2 = \frac{1}{3}$. Also,

$$\sigma_0 = \frac{1}{3}(p_1 + p_2 + p_3) = \frac{I_1}{3} \tag{2.40}$$

and

$$\tau_0 = \frac{1}{3}\sqrt{(p_1 - p_2)^2 + (p_1 - p^3)^2 + (p_2 - p_3)^2} = \frac{1}{3}\sqrt{I_2'}. \tag{2.41}$$

The octahedral stresses can obviously be quoted in terms of the components of any tensor of the state:

$$\sigma_0 = \frac{1}{3}(\sigma_x + \sigma_y + \sigma_z), \tag{2.42}$$

and

$$\tau_0 = \frac{1}{3}\sqrt{(\sigma_x - \sigma_y)^2 + (\sigma_y - \sigma_z)^2 + (\sigma_z - \sigma_x)^2 + 6(\tau_{xy}^2 + \tau_{xz}^2 + \tau_{yz}^2)}. \tag{2.43}$$

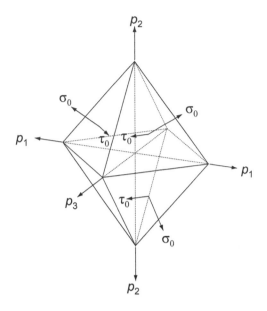

Figure 2.7: State of stress on octahedral sections of an element. $\sigma_0$ is the octahedral normal stress, $\tau_0$ is the octahedral shear stress.

The octahedral shear stress is related to the Von Mises yield stress. Failure occurs when

$$\tau_0 = \frac{\sqrt{2}}{3}\sigma_{ys}. \tag{2.44}$$

## 2.7 Hydrostatic (dilational) and deviatoric stress tensors

It is sometimes necessary to take account of the directions of the octahedral shear stresses as well as their magnitudes. In this section, it is shown that any general stress tensor can be split into two, where one part describes the effect of the direct stress on the octahedral section, whereas the other describes the effect of the octahedral shear stress.

The octahedral normal stress may be expressed in tensor form as

$$[\sigma_0] = \begin{bmatrix} \sigma_0 & 0 & 0 \\ 0 & \sigma_0 & 0 \\ 0 & 0 & \sigma_0 \end{bmatrix}. \tag{2.45}$$

This is called the hydrostatic of dilational stress tensor. It describes a state of stress without shearing present. The value of $\sigma_0$ is the average of the direct stresses on the leading diagonal of the general stress tensor $[\sigma]$. The remainder of the tensor, $[\sigma] - [\sigma_0]$, must describe the effect of the uniform octahedral shears and is called the deviatoric tensor of the state.

Note that the dilational tensor produces a change in volume without change in shape, while the deviatoric tensor produces a change in shape without change in volume.

## 2.8    Strains and displacements

To examine the relationship between strain and displacement, in the first instance only a 2-dimensional case will be considered. The displacements are assumed to be small, so that the strains are small compared with unity.

For the element ABCD in Figure 2.8, stressing the element will result in a displacement of the element in addition to straining. Direct strains are responsible for increases in length of the sides of the element, while shear strains produce rotation of the lines, *i.e.*, change of shape. The increase in length of

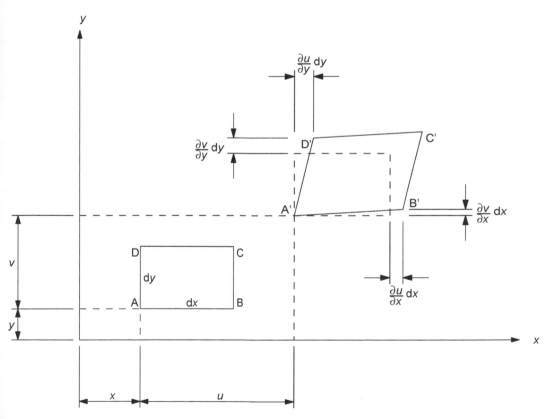

Figure 2.8: Stressing an element.

AB to A'B' is $\mathrm{d}x \times$ (the variation of $u$ with respect to $x$) $= \frac{\partial u}{\partial x}\,\mathrm{d}x$.
Therefore, the direct strain in the $x$-direction is

$$\varepsilon_x = \frac{(\mathrm{d}x + \frac{\partial u}{\partial x}\,\mathrm{d}x) - \mathrm{d}x}{\mathrm{d}x} = \frac{\partial u}{\partial x}. \tag{2.46}$$

Similarly, in the $y$-direction

$$\varepsilon_y = \frac{\partial u}{\partial y}. \tag{2.47}$$

Because of shear, AB and AD rotate through small angles $\theta$ and $\lambda$. For small strains $(\mathrm{d}x \gg \frac{\partial u}{\partial x}\,\mathrm{d}x)$,

$$\theta \approx \tan\theta = \frac{\frac{\partial v}{\partial x}}{\mathrm{d}x + \frac{\partial u}{\partial x}\,\mathrm{d}x} = \frac{\partial v}{\partial x}, \tag{2.48}$$

and

$$\lambda \approx \tan\lambda = \frac{\partial u}{\partial y}. \tag{2.49}$$

The shear strain is the change in angle between two lines originally at right angles:

$$\gamma_{xy} = \theta + \lambda = \frac{\partial v}{\partial x} + \frac{\partial u}{\partial y}. \tag{2.50}$$

Clearly, $\gamma_{yx} = \gamma_{xy}$.

In the case of a 3-dimensional rectangular prism with sides $\mathrm{d}x$, $\mathrm{d}y$, and $\mathrm{d}z$, a similar analysis will give

$$\varepsilon_x = \frac{\partial u}{\partial x}; \qquad \varepsilon_y = \frac{\partial v}{\partial y}; \qquad \varepsilon_z = \frac{\partial w}{\partial z};$$
$$\gamma_{xy} = \frac{\partial u}{\partial y} + \frac{\partial v}{\partial x}; \quad \gamma_{yz} = \frac{\partial v}{\partial z} + \frac{\partial w}{\partial y}; \quad \gamma_{xz} = \frac{\partial w}{\partial x} + \frac{\partial u}{\partial z}. \tag{2.51}$$

Like stress, strain is a tensor quantity. It may be stored in a matrix

$$\begin{bmatrix} \varepsilon_x & \frac{1}{2}\gamma_{yx} & \frac{1}{2}\gamma_{zx} \\ \frac{1}{2}\gamma_{xy} & \varepsilon_y & \frac{1}{2}\gamma_{zy} \\ \frac{1}{2}\gamma_{xz} & \frac{1}{2}\gamma_{yz} & \varepsilon_z \end{bmatrix}. \tag{2.52}$$

A strain tensor has all the properties of a stress tensor, and the same concepts derived in previous Sections will apply to it:

1. Strain resolution

   The normal strain on a plane with direction cosines $l$, $m$, and $n$ is given by

   $$\varepsilon_N = \begin{bmatrix} l & m & n \end{bmatrix} \begin{bmatrix} \varepsilon_x & \frac{1}{2}\gamma_{yx} & \frac{1}{2}\gamma_{zx} \\ \frac{1}{2}\gamma_{xy} & \varepsilon_y & \frac{1}{2}\gamma_{zy} \\ \frac{1}{2}\gamma_{xz} & \frac{1}{2}\gamma_{yz} & \varepsilon_z \end{bmatrix} \begin{bmatrix} l \\ m \\ n \end{bmatrix}. \tag{2.53}$$

2. Transformation of axes

The state of strain $[\varepsilon']$ can be found in any second coordinate system defined by the direction cosine matrix $[L]$ with respect to the first system by the equation

$$[\varepsilon'] = [L][\varepsilon][L]^{\mathsf{T}}. \tag{2.54}$$

3. Principal strains

In any pattern of deformation there is one set of rectangular directions that suffer no relative rotation and are therefore free of shear. The linear strains of these lines are called principal strains. They are found by solving

$$\begin{bmatrix} \varepsilon_x - \varepsilon_p & \frac{1}{2}\gamma_{yx} & \frac{1}{2}\gamma_{zx} \\ \frac{1}{2}\gamma_{xy} & \varepsilon_y - \varepsilon_p & \frac{1}{2}\gamma_{zy} \\ \frac{1}{2}\gamma_{xz} & \frac{1}{2}\gamma_{yz} & \varepsilon_z - \varepsilon_p \end{bmatrix} \begin{bmatrix} l \\ m \\ n \end{bmatrix} = 0. \tag{2.55}$$

In an isotropic material the principal axes of strain will coincide with the principal axes of stress.

4. Stationary values of shear strain

The same technique can be applied to small strains to derive those planes where the maximum and minimum values of the shear strain exist: $\frac{1}{2}\gamma_{13} = \frac{1}{2}(\varepsilon_{p1} - \varepsilon_{p3})$.

5. Volumetric and octahedral strains

The linear strain along each of the four octahedral axes that are equally inclined to the three principal axes is

$$\varepsilon_0 = \frac{1}{3}(\varepsilon_{p1} + \varepsilon_{p2} + \varepsilon_{p3}) = \frac{1}{3}(\varepsilon_x + \varepsilon_y + \varepsilon_z), \tag{2.56}$$

which is related to the volumetric strain $\Delta = \varepsilon_{p1} + \varepsilon_{p2} + \varepsilon_{p3}$. This is proved by considering the volumetric strain of a prism with sides parallel to the principal axes. If the sides are $a_1$, $a_2$, $a_3$ and increase to $a_1 + \varepsilon_{p1}$, $a_2 + \varepsilon_{p2}$, $a_3 + \varepsilon_{p3}$, the volumetric strain

$$\Delta = \frac{\text{change in volume}}{\text{original volume}}$$
$$= \frac{a_1 a_2 a_3 (1 + \varepsilon_{p1})(1 + \varepsilon_{p2})(1 + \varepsilon_{p3}) - a_1 a_2 a_3}{a_1 a_2 a_3}. \tag{2.57}$$

Ignoring second order terms,

$$\Delta = \varepsilon_{p1} + \varepsilon_{p2} + \varepsilon_{p3} = 3\varepsilon_0. \tag{2.58}$$

6. Hydrostatic and deviatoric strain tensors

As for the stress tensor, the strain tensor can be split into its dilational and deviatoric parts. The following relationships apply, which will be discussed in the next Section:

$$\sigma_0[U] = \kappa\Delta[U] \tag{2.59}$$

for the hydrostatic terms, and

$$[\sigma] - \sigma_0[U] = 2G\left([\varepsilon] - \frac{\Delta}{3}[U]\right) \qquad (2.60)$$

for the deviatoric terms. Here, $\kappa$ is the bulk modulus and $G$ is the shear modulus.

## 2.9 Generalised Hooke's law

Hooke's law states that the strain produced in an elastic material is proportional to the applied stress. For a linear elastic material the principle of superposition applied, so that the deformations may be determined independently and added. Remember that a normal stress only produces a normal strain, while a shear stress only produces a shear strain. Therefore, the strain due to $\sigma_x$ in the $x$-direction is $\frac{\sigma_x}{E}$ (extension), with the Poisson effect producing a strain of $-\frac{\nu\sigma_x}{E}$ in the $y$- and $z$-directions. Hence, for a triaxial stress, the total normal strains are given by

$$\varepsilon_x = \frac{\sigma_x}{E} - \frac{\nu\sigma_y}{E} - \frac{\nu\sigma_z}{E} = \frac{1}{E}[\sigma_x - \nu(\sigma_y + \sigma_z)];$$

$$\varepsilon_y = \frac{\sigma_y}{E} - \frac{\nu\sigma_x}{E} - \frac{\nu\sigma_z}{E} = \frac{1}{E}[\sigma_y - \nu(\sigma_z + \sigma_x)]; \qquad (2.61)$$

$$\varepsilon_z = \frac{\sigma_z}{E} - \frac{\nu\sigma_x}{E} - \frac{\nu\sigma_y}{E} = \frac{1}{E}[\sigma_z - \nu(\sigma_x + \sigma_y)],$$

where $E$ is Young's modulus and $\nu$ is Poisson's ratio. Alternatively, in matrix form,

$$\begin{bmatrix} \varepsilon_x \\ \varepsilon_y \\ \varepsilon_z \end{bmatrix} = \frac{1}{E} \begin{bmatrix} 1 & -\nu & -\nu \\ -\nu & 1 & -\nu \\ -\nu & -\nu & 1 \end{bmatrix} \begin{bmatrix} \sigma_x \\ \sigma_y \\ \sigma_z \end{bmatrix}. \qquad (2.62)$$

By solving for the direct stresses we obtain

$$\begin{bmatrix} \sigma_x \\ \sigma_y \\ \sigma_z \end{bmatrix} = \frac{\nu E}{(1-2\nu)(1+\nu)} \begin{bmatrix} \frac{1-\nu}{\nu} & 1 & 1 \\ 1 & \frac{1-\nu}{\nu} & 1 \\ 1 & 1 & \frac{1-\nu}{\nu} \end{bmatrix} \begin{bmatrix} \varepsilon_x \\ \varepsilon_y \\ \varepsilon_z \end{bmatrix}. \qquad (2.63)$$

The shear strains are given by

$$\begin{bmatrix} \gamma_{xy} \\ \gamma_{xz} \\ \gamma_{yz} \end{bmatrix} = \frac{1}{G} \begin{bmatrix} \tau_{xy} \\ \tau_{xz} \\ \tau_{yz} \end{bmatrix} \qquad (2.64)$$

or

$$\begin{bmatrix} \tau_{xy} \\ \tau_{xz} \\ \tau_{yz} \end{bmatrix} = G \begin{bmatrix} \gamma_{xy} \\ \gamma_{xz} \\ \gamma_{yz} \end{bmatrix}. \qquad (2.65)$$

The elastic constants, $E$, $\nu$, and $G$, are related so that there are only two independent constants. To determine the relationship consider a 2-dimensional element in pure shear, $i.e.$, $\sigma_x = \sigma_y = \sigma_z = \tau_{xz} = \tau_{yz} = 0$ and $\tau_{xy} \neq 0$.

The principal stresses for this system, acting on planes at $45°$ to the $xy$-axes, are

$$\begin{aligned} p_1 &= \tau_{xy} \; ; \\ p_2 &= -\tau_{xy} \; ; \\ p_3 &= 0 \; . \end{aligned} \tag{2.66}$$

From Hooke's law,

$$\varepsilon_{p_1} = \frac{p_1}{E} - \frac{\nu p_2}{E} = \frac{\tau_{xy}}{E}(1 + \nu). \tag{2.67}$$

If the strains are resolved into a direction at $45°$ to the $xy$-axes, $l = \cos\theta = m = \sin\theta = \frac{1}{\sqrt{2}}$. Combining this with (2.64) gives

$$\varepsilon_{p_1} = \frac{1}{2}\gamma_{xy} = \frac{\tau_{xy}}{2G}. \tag{2.68}$$

From (2.67) it now follows that

$$G = \frac{E}{2(1 + \nu)}. \tag{2.69}$$

## 2.9.1 The bulk modulus

Summing the three equations in (2.61) gives

$$\varepsilon_x + \varepsilon_y + \varepsilon_z = \frac{1 - 2\nu}{E}(\sigma_x + \sigma_y + \sigma_z). \tag{2.70}$$

The sum on the left is the invariant that determines the volumetric strain (2.58), while the right-hand term is the invariant that is equal to three times the hydrostatic stress (2.42). Therefore,

$$\Delta = \frac{3(1 - 2\nu)}{E}\sigma_0 = \frac{\sigma_0}{\kappa}, \tag{2.71}$$

where $\kappa = \frac{E}{3(1-2\nu)}$ is an elastic constant called the bulk modulus.

The relationship between the deviatoric tensors can easily be determined. From (2.63), by subtracting the second and third equations from twice the first, we obtain

$$\sigma_x - \sigma_0 = \frac{E}{1 + \nu}\left(\varepsilon_x - \frac{\Delta}{3}\right) = 2G\left(\varepsilon_x - \frac{\Delta}{3}\right). \tag{2.72}$$

Thus, each direct stress of the deviatoric tensor, like each shear component, is related by $2G$ to the corresponding element of the deviatoric strain tensor.

## 2.9.2 Lamé's constant

It has been shown that the total stress tensor can be split into its hydrostatic and deviatoric parts. Each of these effects may then be described as a simple proportion of the corresponding strain effect. This is advantageous since we can often ignore the hydrostatic effect. In other cases, however, it may be more convenient to retain the total tensors intact. Here we show that the split relationship can be recombined to describe the total stress tensor as a linear function of the total strain tensor and the volumetric strain:

$$[\sigma] = 2G[\varepsilon] + \left(\kappa - \frac{2G}{3}\right)\Delta[U]. \tag{2.73}$$

Because $\kappa$ and $G$ are constants,

$$[\sigma] = 2G[\varepsilon] + \lambda\Delta[U], \tag{2.74}$$

where $\lambda$ is Lamé's constant:

$$\lambda = \kappa - \frac{2G}{3} = \frac{E\nu}{(1+\nu)(1-2\nu)}. \tag{2.75}$$

# 2.10 Equilibrium equations for three dimensions

Consider a small cuboid of finite size $\Delta x \times \Delta y \times \Delta z$, with stresses acting on each coordinate plane and their variations on opposite faces. Resolving the forces in the $x$-direction and equating to zero for equilibrium, ignoring body forces, yields

$$\sigma_x\,\Delta z\,\Delta y - \left(\sigma_x + \frac{\partial\sigma_x}{\partial x}\Delta x\right)\Delta z\,\Delta y$$

$$+\tau_{yx}\,\Delta x\,\Delta z - \left(\tau_{yx} + \frac{\partial\tau_{yx}}{\partial y}\Delta y\right)\Delta x\,\Delta z \tag{2.76}$$

$$+\tau_{zx}\,\Delta x\,\Delta y - \left(\tau_{zx} + \frac{\partial\tau_{zx}}{\partial z}\Delta z\right)\Delta x\,\Delta y = 0,$$

which after simplifying becomes

$$\frac{\partial\sigma_x}{\partial x} + \frac{\partial\tau_{yx}}{\partial y} + \frac{\partial\tau_{zx}}{\partial z} = 0. \tag{2.77}$$

Resolving in the other directions similarly gives

$$[\sigma]\begin{bmatrix} \frac{\partial}{\partial x} \\ \frac{\partial}{\partial y} \\ \frac{\partial}{\partial z} \end{bmatrix} = \begin{bmatrix} 0 \\ 0 \\ 0 \end{bmatrix}. \tag{2.78}$$

These equations ensure that equilibrium of the material is maintained.

In two dimensions they can be simplified to

$$\frac{\partial \sigma_x}{\partial x} + \frac{\partial \tau_{yx}}{\partial y} = 0 \; ;$$

$$\frac{\partial \tau_{xy}}{\partial x} + \frac{\partial \sigma_y}{\partial y} = 0 \; . \tag{2.79}$$

## 2.11  Strain compatibility equations

The fundamental equations (2.51) relate the six components of strain in a 3-dimensional system to only three components of displacement. The strains therefore cannot be independent of one another.

Consider a 2-dimensional case. Since

$$\varepsilon_x = \frac{\partial u}{\partial x} \tag{2.80}$$

and

$$\varepsilon_y = \frac{\partial v}{\partial y}, \tag{2.81}$$

it follows that

$$\frac{\partial^2 \varepsilon_x}{\partial y^2} = \frac{\partial^3 u}{\partial x \partial y^2} \tag{2.82}$$

and

$$\frac{\partial^2 \varepsilon_y}{\partial x^2} = \frac{\partial^3 v}{\partial x^2 \partial y}. \tag{2.83}$$

Also,

$$\frac{\partial^2 \gamma_{xy}}{\partial x \partial y} = \frac{\partial^3 u}{\partial x \partial y^2} + \frac{\partial^3 v}{\partial x^2 \partial y}. \tag{2.84}$$

Therefore,

$$\frac{\partial^2 \varepsilon_x}{\partial y^2} + \frac{\partial^2 \varepsilon_y}{\partial x^2} = \frac{\partial^2 \gamma_{xy}}{\partial x \partial y}. \tag{2.85}$$

This is the condition of compatibility for 2-dimensional problems. The 3-dimensional equations of compatibility are derived in a similar manner:

$$\frac{\partial^2 \varepsilon_x}{\partial y^2} + \frac{\partial^2 \varepsilon_y}{\partial x^2} = \frac{\partial^2 \gamma_{xy}}{\partial x \partial y};$$

$$\frac{\partial^2 \varepsilon_y}{\partial z^2} + \frac{\partial^2 \varepsilon_z}{\partial y^2} = \frac{\partial^2 \gamma_{yz}}{\partial y \partial z}; \tag{2.86}$$

$$\frac{\partial^2 \varepsilon_z}{\partial x^2} + \frac{\partial^2 \varepsilon_x}{\partial z^2} = \frac{\partial^2 \gamma_{xz}}{\partial x \partial z}$$

and

$$2\frac{\partial^2 \varepsilon_x}{\partial y \partial z} = \frac{\partial}{\partial x}\left(-\frac{\partial \gamma_{yz}}{\partial x} + \frac{\partial \gamma_{xz}}{\partial y} + \frac{\partial \gamma_{xy}}{\partial z}\right);$$

$$2\frac{\partial^2 \varepsilon_y}{\partial z \partial x} = \frac{\partial}{\partial y}\left(\frac{\partial \gamma_{yz}}{\partial x} - \frac{\partial \gamma_{xz}}{\partial y} + \frac{\partial \gamma_{xy}}{\partial z}\right); \qquad (2.87)$$

$$2\frac{\partial^2 \varepsilon_x}{\partial x \partial y} = \frac{\partial}{\partial z}\left(\frac{\partial \gamma_{yz}}{\partial x} + \frac{\partial \gamma_{xz}}{\partial y} - \frac{\partial \gamma_{xy}}{\partial z}\right).$$

## 2.12  Plane strain

Consider a long cylinder held between fixed, rigid end plates. It is subjected to an internal pressure, *i.e.*, a purely lateral load. The cylinder can only deform in the $x$- and $y$-directions. So, $w = 0$ along the length of the cylinder. From (2.51),

$$\varepsilon_z = \gamma_{xz} = \gamma_{yz} = 0 \qquad (2.88)$$

and

$$\varepsilon_x = \frac{\partial u}{\partial x};$$

$$\varepsilon_y = \frac{\partial v}{\partial y}; \qquad (2.89)$$

$$\gamma_{xy} = \frac{\partial u}{\partial y} + \frac{\partial v}{\partial x}.$$

This is a state of plain strain, where each point remains within its transverse plane following application of the load. Since $\varepsilon_z = 0$,

$$\varepsilon_z = \frac{1}{E}\left[\sigma_x - \nu(\sigma_x + \sigma_y)\right] = 0, \qquad (2.90)$$

so that

$$\varepsilon_x = \frac{1-\nu^2}{E}\left(\sigma_x - \frac{\nu}{1-\nu}\sigma_y\right);$$

$$\varepsilon_y = \frac{1-\nu^2}{E}\left(\sigma_y - \frac{\nu}{1-\nu}\sigma_x\right); \qquad (2.91)$$

$$\gamma_{xy} = \frac{1}{G}\tau_{xy}.$$

The compatibility equation must also be satisfied by this stressing regime, for two dimensions that is:

$$\frac{\partial^2 \varepsilon_x}{\partial y^2} + \frac{\partial^2 \varepsilon_y}{\partial x^2} = \frac{\partial^2 \gamma_{xy}}{\partial x \partial y}. \qquad (2.92)$$

Differentiating (2.91) and using (2.79) with (2.92) gives

$$\left(\frac{\partial^2}{\partial x^2} + \frac{\partial^2}{\partial y^2}\right)(\sigma_x + \sigma_y) = 0. \qquad (2.93)$$

This is the equation of compatibility in terms of stress. Therefore, we now have three equations defining three unknowns:

$$\frac{\partial \sigma_x}{\partial x} + \frac{\partial \tau_{yx}}{\partial y} = 0 \; ;$$

$$\frac{\partial \tau_{xy}}{\partial x} + \frac{\partial \sigma_y}{\partial y} = 0 \; ; \qquad (2.94)$$

$$\nabla^2 (\sigma_x + \sigma_y) = 0 \; .$$

## 2.13 Plane stress

Consider a thin plate whose loading is evenly distributed over the thickness parallel to the plane of the plate. On the faces of the plate, $\sigma_z = \tau_{zx} = \tau_{zy} = 0$. Since the plate is thin, it can be assumed the same is true through its thickness. The stress–strain relationships are therefore

$$\varepsilon_x = \frac{1}{E} (\sigma_x - \nu \sigma_y) \; ;$$

$$\varepsilon_y = \frac{1}{E} (\sigma_y - \nu \sigma_z) \; ;$$

$$\varepsilon_z = -\frac{\nu}{E} (\sigma_x + \sigma_y) \; ; \qquad (2.95)$$

$$\gamma_{xy} = \frac{1}{G} \tau_{xy} \; .$$

Again, using the same strain compatibility equation with the equations for equilibrium, it can be proven that

$$\nabla^2 (\sigma_x + \sigma_y) = 0, \qquad (2.96)$$

which is the same equation of compatibility as for the plane strain case.

## 2.14 Polar coordinates

Where axial symmetry exists, it is much more convenient to use polar coordinates. The polar coordinate system $(r, \theta)$ and the Cartesian coordinate system $(x, y)$ are related by the expressions

$$x = r \cos \theta;$$
$$y = r \sin \theta;$$
$$r^2 = x^2 + y^2; \qquad (2.97)$$
$$\theta = \arctan \frac{y}{x}.$$

### 2.14.1   Strain components in polar coordinates

Consider a pure radial strain, where the internal face of the element with original length $dr$ is displaced by a radial distance $u$ and a tangential distance $v$. The strained length of the element is $dr + \frac{\partial u}{\partial r} dr$. Therefore, the radial strain

$$\varepsilon_r = \frac{\frac{\partial u}{\partial r} dr}{dr} = \frac{\partial u}{\partial r}. \tag{2.98}$$

The tangential strain arises from the radial displacement

$$\varepsilon_{\theta_r} = \frac{(r + u)\, d\theta - r\, d\theta}{r\, d\theta} = \frac{u}{r} \tag{2.99}$$

and the tangential displacement

$$\varepsilon_{\theta_\theta} = \frac{\left(r + \frac{\partial v}{\partial \theta}\right) d\theta - r\, d\theta}{r\, d\theta} = \frac{1}{r}\frac{\partial v}{\partial \theta}. \tag{2.100}$$

Thus,

$$\varepsilon_\theta = \frac{u}{r} + \frac{1}{r}\frac{\partial v}{\partial \theta}. \tag{2.101}$$

The shear strain also has two components arising from each of the $u$ and $v$ components. The total shear strain is

$$\gamma_{r\theta} = \frac{\partial v}{\partial r} + \frac{1}{r}\frac{\partial u}{\partial \theta} - \frac{v}{r}. \tag{2.102}$$

So, in two dimensions, the strain components are

$$\begin{aligned}
\varepsilon_r &= \frac{\partial u}{\partial r}; \\[6pt]
\varepsilon_\theta &= \frac{u}{r} + \frac{1}{r}\frac{\partial v}{\partial \theta}; \\[6pt]
\gamma_{r\theta} &= \frac{\partial v}{\partial r} + \frac{1}{r}\frac{\partial u}{\partial \theta} - \frac{v}{r}.
\end{aligned} \tag{2.103}$$

### 2.14.2   Hooke's law

To write Hooke's law in polar coordinates, the $x$ and $y$ subscripts are simply replaced by $r$ and $\theta$. For example, in plane stress:

$$\begin{aligned}
\varepsilon_r &= \frac{1}{E}\left(\sigma_r - \nu\sigma_\theta\right); \\[6pt]
\varepsilon_\theta &= \frac{1}{E}\left(\sigma_\theta - \nu\sigma_r\right); \\[6pt]
\gamma_{r\theta} &= \frac{1}{G}\,\tau_{r\theta}.
\end{aligned} \tag{2.104}$$

### 2.14.3 Equilibrium equations

By considering equilibrium of the radial and circumferential forces acting on an element of unit thickness, we obtain

$$\frac{r \partial \sigma_r}{\partial r} + \frac{\partial \tau_{r\theta}}{\partial \theta} + \sigma_r - \sigma_\theta = 0 \; ;$$

$$\frac{\partial \sigma_\theta}{\partial \theta} + \frac{r \partial \tau_{r\theta}}{\partial r} + 2\tau_{r\theta} = 0 \; . \tag{2.105}$$

### 2.14.4 Strain compatibility equation

The three compatibility equations (2.103) defining the strain components can be combined to give the equation of compatibility

$$\frac{\partial^2 \varepsilon_\theta}{\partial r^2} + \frac{1}{r^2} \frac{\partial^2 \varepsilon_r}{\partial \theta^2} + \frac{2}{r} \frac{\partial \varepsilon_\theta}{\partial r} - \frac{1}{r} \frac{\partial \varepsilon_r}{\partial r} = \frac{1}{r} \frac{\partial^2 \gamma_{r\theta}}{\partial r \partial \theta} + \frac{1}{r^2} \frac{\partial \gamma_{r\theta}}{\partial \theta}. \tag{2.106}$$

### 2.14.5 Stress compatibility equation

Substituting (2.104) into (2.106) yields

$$\left( \frac{\partial^2}{\partial r^2} + \frac{1}{r} \frac{\partial}{\partial r} + \frac{1}{r^2} \frac{\partial^2}{\partial \theta^2} \right) (\sigma_r + \sigma_\theta) = 0. \tag{2.107}$$

For axially symmetrical problems, this may be further simplified to

$$\left( \frac{\partial^2}{\partial r^2} + \frac{1}{r} \frac{\partial}{\partial r} \right) (\sigma_r + \sigma_\theta) = 0. \tag{2.108}$$

## 2.15 Stress functions

The preceding theory has shown that the solution of 2-dimensional problems in elasticity requires the integration of the differential equations of equilibrium together with the compatibility equations and the boundary conditions for the problem.

In 1862, G.B. Airy proposed a stress function $\phi(x, y)$ that satisfied the above conditions, defined by

$$\sigma_x = \frac{\partial^2 \phi}{\partial y^2} ;$$

$$\sigma_y = \frac{\partial^2 \phi}{\partial x^2} ; \tag{2.109}$$

$$\tau_{xy} = -\frac{\partial^2 \phi}{\partial x \partial y}.$$

When substituted into (2.96), we obtain

$$\left(\frac{\partial^2}{\partial x^2} + \frac{\partial^2}{\partial y^2}\right)\left(\frac{\partial^2\phi}{\partial y^2} + \frac{\partial^2\phi}{\partial x^2}\right) = 0 \tag{2.110}$$

or

$$\frac{\partial^4\phi}{\partial x^4} + 2\frac{\partial^4\phi}{\partial x^2\partial y^2} + \frac{\partial^4\phi}{\partial y^4} = 0, \tag{2.111}$$

which is equal to

$$\nabla^4\phi = 0. \tag{2.112}$$

This is a biharmonic differential equation representing the compatibility equation for stress. It is difficult to solve for any but the simplest of problems. The advantage of the Airy stress function, however, is that functions of $x$ and $y$ can be devised that represent particular stressing regimes. This is known as the semi-inverse method of analysis, while developing $\phi$ from known boundary conditions is the direct method.

## Examples

1. Single powers of $x$ and $y$ are no use because they give $\sigma_x = \sigma_y = \tau_{xy} = 0$.

2. $\phi = Ax^2$ represents simple tension in the $x$-direction: $\sigma_x = 0$; $\sigma_y = 2A$; $\tau_{xy} = 0$.

3. $\phi = Axy$ represents pure shear parallel to the axes: $\sigma_x = \sigma_y = 0$; $\tau_{xy} = -A$.

4. $\phi = Ay^3$ represents pure bending: $\sigma_x = 6Ay$; $\sigma_y = 0$; $\tau_{xy} = 0$.

5. $\phi = Ax^4$ does not satisfy (2.112).

6. $\phi = A\left(xy^3 - \frac{3}{4}xyh^2\right)$ represents a thin cantilever of thickness $h$, end-loaded by force $F$: $\sigma_x = 6Axy$; $\sigma_y = 0$; $\tau_{xy} = 3A\left(\frac{1}{4}h^2 - y^2\right)$.
   The boundary conditions are satisfied: $\tau_{xy} = 0$ at $y = \pm\frac{1}{2}h$ for all values of $x$; $\sigma_y = 0$ at $y = \pm\frac{1}{2}h$ for all values of $x$; $\sigma_x = 0$ at $x = 0$ for all values of $y$.
   The magnitude of $F$ can be found by different approaches, for example:
   $$P = 2\int_0^{\frac{h}{2}} \tau_{xy}b\,dy \text{ for beam width } b, \text{ giving } F = \frac{Ah^3b}{2}.$$

## 2.16   Stress functions in polar coordinates

The Airy stress function may also be used in polar coordinates:

$$\sigma_r = \frac{1}{r^2}\frac{\partial^2 \phi}{\partial \theta^2} + \frac{1}{r}\frac{\partial \phi}{\partial r};$$

$$\sigma_\theta = \frac{\partial^2 \phi}{\partial r^2};$$

$$\tau_{r\theta} = -\frac{\partial}{\partial r}\left(\frac{1}{r}\frac{\partial \phi}{\partial \theta}\right).$$

(2.113)

The biharmonic equation then becomes

$$\nabla^4 \phi = \left(\frac{\partial^2}{\partial r^2} + \frac{1}{r}\frac{\partial}{\partial r} + \frac{1}{r^2}\frac{\partial^2}{\partial \theta^2}\right)\left(\frac{\partial^2 \phi}{\partial r^2} + \frac{1}{r}\frac{\partial \phi}{\partial r} + \frac{1}{r^2}\frac{\partial^2 \phi}{\partial \theta^2}\right) = 0. \qquad (2.114)$$

# 3

## Vibrations

with Keith Attenborough

Before we can continue to the propagation of sound waves, we have to treat the principles of free and forced vibrations.

## 3.1   Mass on a spring

Consider an object of a mass $m$ attached by a massless elastic spring to a motionless ceiling at rest, as displayed in Figure 3.1. Let us define this equilibrium position as $x = 0$. The mass is pulled back over a distance $\Delta x$ by a force $F$. For a stationary mass,

$$F = s\,\Delta x, \tag{3.1}$$

where $s$ is the stiffness of the spring. The elastic potential energy gained equals the work done by stretching the elastic object:

$$E_\mathrm{P} = \int_0^{\Delta x} F\,\mathrm{d}x = \int_0^{\Delta x} s\,x\,\mathrm{d}x = \left[\frac{1}{2}\,s\,x^2\right]_0^{\Delta x} = \frac{1}{2}\,s\,\Delta x^2. \tag{3.2}$$

After release, we can define the excursion $x(t)$ around the equilibrium by

$$m\,\ddot{x} = -s\,x \tag{3.3}$$

or

$$a = -\frac{s}{m}\,x. \tag{3.4}$$

Using

$$a = \frac{\mathrm{d}v}{\mathrm{d}t} = \frac{\mathrm{d}v}{\mathrm{d}x}\frac{\mathrm{d}x}{\mathrm{d}t} = v\frac{\mathrm{d}v}{\mathrm{d}x}, \tag{3.5}$$

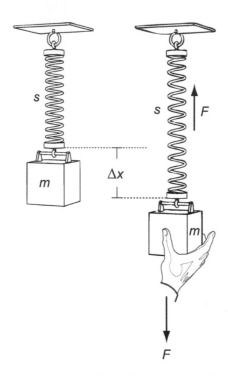

Figure 3.1: Mass $m$ on a spring with stiffness $s$, pulled back by a distance $\Delta x$.

we get

$$v\frac{\mathrm{d}v}{\mathrm{d}x} = -\frac{s}{m}x. \tag{3.6}$$

This can be rearranged to

$$\int v\,\mathrm{d}v = \int -\frac{s}{m}x\,\mathrm{d}x \tag{3.7}$$

or

$$\frac{v^2}{2} = -\frac{s}{m}\frac{x^2}{2} + c. \tag{3.8}$$

Since $v(0) = 0$ and $x(0) = \Delta x$, $c = \frac{s}{m}(\Delta x)^2$, yielding

$$v = \frac{\mathrm{d}x}{\mathrm{d}t} = \sqrt{\frac{s}{m}(\Delta x)^2 - \frac{s}{m}x^2}, \tag{3.9}$$

from which it follows that

$$\sqrt{\frac{m}{s}}\int \frac{1}{\sqrt{(\Delta x)^2 - x^2}}\,\mathrm{d}x = \int \mathrm{d}t. \tag{3.10}$$

Using $\frac{d}{dx} \arcsin x = \frac{1}{\sqrt{1-x^2}}$, this becomes

$$\sqrt{\frac{m}{s}} \arcsin \frac{x}{\Delta x} = t + c \tag{3.11}$$

or

$$\arcsin \frac{x}{\Delta x} = \sqrt{\frac{s}{m}} (t + c). \tag{3.12}$$

Hence, a solution for (3.3) is given by

$$x = \Delta x \sin \left( \sqrt{\frac{s}{m}} (t + c) \right). \tag{3.13}$$

This can be further simplified to

$$x = \Delta x \sin (\omega_0 t + \phi), \tag{3.14}$$

where

$$\omega_0 = \sqrt{\frac{s}{m}} \tag{3.15}$$

is the natural (or resonance) frequency of the system in radians per unit time, and

$$\omega_0 = 2\pi f_0 = \frac{2\pi}{T_0}, \tag{3.16}$$

where $f_0$ is the resonance frequency in cycles per unit time and $T_0$ is the resonance period. The excursion is maximal at $x = 0$, the phase $\phi = \frac{1}{2}\pi$. Thus, (3.14) becomes

$$x = \Delta x \cos \omega_0 t. \tag{3.17}$$

## 3.2 Free vibrations

Consider an undamped mass $m$ on a spring with stiffness $s$, (3.3), the same as displayed in Figure 3.1:

$$m\ddot{x} + sx = 0 \tag{3.18}$$

or

$$\ddot{x} + \omega_0^2 x = 0, \tag{3.19}$$

where $\omega_0 = \sqrt{\frac{s}{m}}$ is the resonance frequency of the system (3.15). The general solution of this ordinary differential equation is

$$x = A \cos \omega_0 t + B \sin \omega_0 t. \tag{3.20}$$

If $x(0) = x_0$ and $\dot{x}(0) = \dot{x}_0$,

$$\begin{aligned}
x &= x_0 \cos \omega_0 t + \frac{\dot{x}_0}{\omega_0} \sin \omega_0 t \\
&= C \sin (\omega_0 t + \phi) \\
&= C (\sin \omega_0 t \cos \phi + \cos \omega_0 t \sin \phi) \\
&= C \sin \phi \cos \omega_0 t + C \cos \phi \sin \omega_0 t.
\end{aligned} \tag{3.21}$$

Since $C \sin \phi = x_0$ and $C \cos \phi = \frac{\dot{x}_0}{\omega_0}$,

$$x = \sqrt{x_0^2 + \frac{\dot{x}^2}{\omega_0^2}} \, \sin \left( \omega_0 t + \arctan \frac{x_0 \, \omega_0}{\dot{x}_0} \right). \tag{3.22}$$

## 3.3   Damped free vibrations

Consider a mass–spring–dashpot system, for which the damping force is proportional to the velocity, as demonstrated in Figure 3.2. Then (3.18) becomes

$$m \, \ddot{x} + \beta \dot{x} + s \, x = 0, \tag{3.23}$$

where $\beta$ is the mechanical resistance. Using (3.15) and the damping coefficient

$$\zeta = \frac{\beta}{2m\omega_0}, \tag{3.24}$$

we can express the system in terms of its damping and resonance frequency:

$$\ddot{x} + 2\zeta\omega_0 \dot{x} + \omega_0^2 x = 0. \tag{3.25}$$

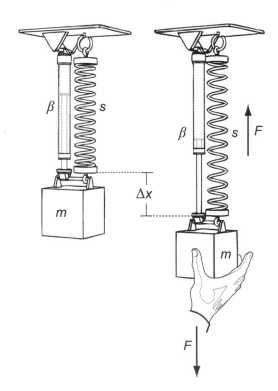

Figure 3.2: Mass $m$ on a spring with stiffness $s$, pulled back by a distance $\Delta x$, resisted by a dashpot with resistance $\beta$.

Assume $x = Ae^{\lambda t}$. Then,

$$\lambda^2 + 2\zeta\omega_0\lambda + \omega_0^2 = 0. \tag{3.26}$$

Thus,

$$\lambda = \frac{-2\zeta\omega_0 \pm \sqrt{4\zeta^2\omega_0^2 - 4\omega_0^2}}{2} = \omega_0\left(-\zeta \pm \sqrt{\zeta^2 - 1}\right) \tag{3.27}$$

and

$$x = A_1 e^{(-\zeta+\sqrt{\zeta^2-1})\omega_0 t} + A_2 e^{(-\zeta-\sqrt{\zeta^2-1})\omega_0 t}. \tag{3.28}$$

Now, different forms of motion result from different $\zeta$ values and this is the reason for the assumed form of $\zeta$. The following three conditions of damping exist, represented in Figure 3.3:

1. Overdamped condition $\zeta > 1$
   Since $\sqrt{\zeta^2 - 1}$ is real, (3.26) has two distinct roots.

2. Critically damped condition $\zeta = 1$
   There is only one root: $\lambda = -\omega_0$. Therefore,

$$x = Ae^{-\omega_0 t}. \tag{3.29}$$

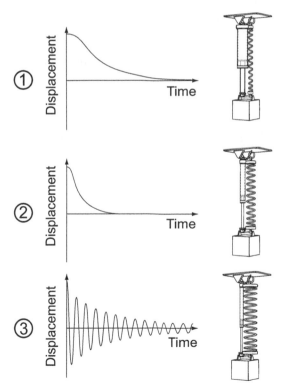

Figure 3.3: Schematic representation of (1) overdamped, (2) critically damped, and (3) underdamped conditions.

3. Underdamped condition $\zeta < 1$

Since $\sqrt{\zeta^2 - 1}$ is imaginary, the solution is complex:

$$x = A_1 e^{(-\zeta + j\sqrt{1-\zeta^2})\omega_0 t} + A_2 e^{(-\zeta - j\sqrt{1-\zeta^2})\omega_0 t}$$

$$= \left( A_1 e^{j\omega_0 t \sqrt{1-\zeta^2}} + A_2 e^{-j\omega_0 t \sqrt{1-\zeta^2}} \right) e^{-\zeta\omega_0 t}. \tag{3.30}$$

Making use of $e^{j\theta} = \cos\theta + j\sin\theta$, we can reformulate this as

$$\begin{aligned}
x &= \left( A_1 e^{j\omega_d t} + A_2 e^{-j\omega_d t} \right) e^{-\zeta\omega_0 t} \\
&= [A_1 (\cos\omega_d t + j\sin\omega_d t) + A_2 (\cos\omega_d t - j\sin\omega_d t)] e^{-\zeta\omega_0 t} \\
&= [(A_1 + A_2)\cos\omega_d t + j(A_1 - A_2)\sin\omega_d t] e^{-\zeta\omega_0 t} \tag{3.31} \\
&= [B_1 \cos\omega_d t + B_2 \sin\omega_d t] e^{-\zeta\omega_0 t} \\
&= C e^{-\zeta\omega_0 t} \sin(\omega_d t + \phi),
\end{aligned}$$

where $\omega_d$ is the damped natural frequency

$$\omega_d = \omega_0 \sqrt{1 - \zeta^2}. \tag{3.32}$$

So the excursion amplitude decays by $e^{-\zeta\omega_0 t}$. The damped period is

$$T_d = \frac{2\pi}{\omega_d}. \tag{3.33}$$

The damping coefficient can be deduced by measuring the excursion amplitudes $x_1 = x(t_1)$ and $x_2 = x(t_2)$ at two times whose difference $t_2 - t_1 = T_d$:

$$\delta \equiv \log_e \frac{x_1}{x_2} = \zeta\omega_0 T_d = \frac{2\pi\zeta}{\sqrt{1 - \zeta^2}}. \tag{3.34}$$

Thus,

$$\zeta = \frac{\delta}{\sqrt{(2\pi)^2 + \delta^2}}. \tag{3.35}$$

## 3.4 Forced vibrations

Consider a mass–spring–dashpot system whose mass is subjected to a periodic force $F$ with amplitude $F_0$:

$$F = F_0 \sin\omega t, \tag{3.36}$$

as shown in Figure 3.4. The excursion of the mass can be described by adding $F$ to (3.23):

$$-s\,x - \beta\dot{x} + F_0 \sin\omega t = m\,\ddot{x}. \tag{3.37}$$

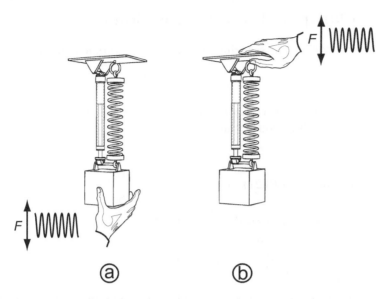

Figure 3.4: Schematic representation of (a) a mass-forced system and (b) a base-forced system.

Using (3.15) and (3.24), we can express the mass-forced system in terms of its damping and resonance frequency:

$$\ddot{x} + 2\zeta\omega_0\dot{x} + \omega_0^2 x = \frac{F_0 \sin\omega t}{m}. \tag{3.38}$$

Now consider a mass–spring–dashpot system whose end is subjected to a periodic excursion $x_b$ with amplitude $b$:

$$x_b = b \sin\omega t. \tag{3.39}$$

The excursion of the mass can be described by

$$-s\,(x - x_b) - \beta\dot{x} = m\,\ddot{x} \tag{3.40}$$

or

$$\ddot{x} + \frac{\beta}{m}\,\dot{x} + \frac{s}{m}\,x = \frac{s\,b\,\sin\omega t}{m}. \tag{3.41}$$

A base-forced system can be expressed in terms of its damping and resonance frequency by

$$\ddot{x} + 2\zeta\omega_0\dot{x} + \omega_0^2 x = \frac{s\,b\,\sin\omega t}{m}. \tag{3.42}$$

## 3.5 Undamped forced vibrations

Consider a mass-forced mass–spring system. If the system is undamped, (3.38) reduces to

$$\ddot{x} + \omega_0^2 x = \frac{F_0 \sin \omega t}{m}.$$

(3.43)

The solution consists of the complementary function and the particular integral. The complementary function is the solution of

$$\ddot{x} + \omega_0^2 x = 0,$$

(3.44)

which has been shown in (3.21) and (3.22):

$$x = x_0 \cos \omega_0 t + \frac{\dot{x}_0}{\omega_0} \sin \omega_0 t$$

$$= \sqrt{x_0^2 + \frac{\dot{x}^2}{\omega_0^2}} \sin \left( \omega_0 t + \arctan \frac{x_0 \omega_0}{\dot{x}_0} \right).$$

(3.45)

This is called the transient solution. The particular integral represents the steady-state solution. It can be determined assuming the same form as the driving function

$$x = X \sin \omega t.$$

(3.46)

Hence, substituting in (3.43) gives

$$-\omega^2 X \sin \omega t + \omega_0^2 X \sin \omega t = \frac{F_0 \sin \omega t}{m}.$$

(3.47)

Dividing through by $\omega_0^2 \sin \omega t$ and rearranging results in

$$\left[ 1 - \left( \frac{\omega}{\omega_0} \right)^2 \right] X = \frac{F_0}{m \omega_0^2} = \frac{F_0}{s}$$

(3.48)

or

$$X = \frac{\dfrac{F_0}{s}}{1 - \left( \dfrac{\omega}{\omega_0} \right)^2}.$$

(3.49)

So the steady-state solution of (3.43) is

$$x = \frac{\dfrac{F_0}{s}}{1 - \left( \dfrac{\omega}{\omega_0} \right)^2} \sin \omega t.$$

(3.50)

The static deflection of a mass under a static load $F_0$ is denoted by

$$\delta_s = \frac{F_0}{s}.$$

(3.51)

Now, the severity of the vibration can be expressed in terms of the amplitude ratio or magnification factor

$$M = \frac{X}{\delta_s} = \frac{1}{1 - \left(\dfrac{\omega}{\omega_0}\right)^2}. \tag{3.52}$$

## 3.6 Damped forced vibrations

Consider a damped mass-forced mass–spring–dashpot system. The excursion around equilibrium is given by (3.38):

$$\ddot{x} + 2\zeta\omega_0\dot{x} + \omega_0^2 x = \frac{F_0 \sin \omega t}{m}. \tag{3.53}$$

The complementary function is a solution of

$$\ddot{x} + 2\zeta\omega_0\dot{x} + \omega_0^2 x = 0, \tag{3.54}$$

which has been shown in (3.31):

$$x = Ce^{-\zeta\omega_0 t} \sin(\omega_d t + \phi). \tag{3.55}$$

The transient part of the solution damps out. The particular integral can be determined assuming the form

$$x = A \cos \omega t + B \sin \omega t = X \sin(\omega t - \phi). \tag{3.56}$$

Substituting in (3.53) gives

$$\begin{aligned} -\omega^2 X \sin(\omega t - \phi) + 2\zeta\omega_0 \omega X \cos(\omega t - \phi) \\ +\omega_0^2 X \sin(\omega t - \phi) &= \frac{F_0 \sin \omega t}{m}. \end{aligned} \tag{3.57}$$

Dividing through by $\omega_0^2$ and rearranging results in

$$\left[1 - \left(\frac{\omega}{\omega_0}\right)^2\right] X \sin(\omega t - \phi) + 2\zeta\frac{\omega}{\omega_0} X \cos(\omega t - \phi) = \frac{F_0 \sin \omega t}{m\,\omega_0^2}. \tag{3.58}$$

This can be rewritten as

$$\begin{aligned} \left[1 - \left(\frac{\omega}{\omega_0}\right)^2\right] X (\sin \omega t \, \cos \phi - \cos \omega t \, \sin \phi) \\ +2\zeta\frac{\omega}{\omega_0} X (\cos \omega t \, \cos \phi + \sin \omega t \, \sin \phi) &= \frac{F_0 \sin \omega t}{m\,\omega_0^2} \end{aligned} \tag{3.59}$$

or

$$\left\{ -\left[1 - \left(\frac{\omega}{\omega_0}\right)^2\right] X \sin\phi + 2\zeta \frac{\omega}{\omega_0} X \cos\phi \right\} \cos\omega t$$
$$+ \left\{ \left[1 - \left(\frac{\omega}{\omega_0}\right)^2\right] X \cos\phi + 2\zeta \frac{\omega}{\omega_0} X \sin\phi \right\} \sin\omega t = \frac{F_0 \sin\omega t}{m\,\omega_0^2}. \tag{3.60}$$

This can be split up into two equations:

$$\left\{ -\left[1 - \left(\frac{\omega}{\omega_0}\right)^2\right] X \sin\phi + 2\zeta \frac{\omega}{\omega_0} X \cos\phi \right\} \cos\omega t = 0 \tag{3.61}$$

and

$$\left\{ \left[1 - \left(\frac{\omega}{\omega_0}\right)^2\right] X \cos\phi + 2\zeta \frac{\omega}{\omega_0} X \sin\phi \right\} \sin\omega t = \frac{F_0 \sin\omega t}{m\,\omega_0^2}. \tag{3.62}$$

From (3.61), it follows that

$$\phi = \arctan\left( \frac{2\zeta \dfrac{\omega}{\omega_0}}{1 - \left(\dfrac{\omega}{\omega_0}\right)^2} \right). \tag{3.63}$$

After dividing through by $\sin\omega t$, (3.62) is simplified to

$$\left[1 - \left(\frac{\omega}{\omega_0}\right)^2\right] X \cos\phi + 2\zeta \frac{\omega}{\omega_0} X \sin\phi = \frac{F_0}{m\,\omega_0^2} = \frac{F_0}{s}. \tag{3.64}$$

Combining (3.63) and (3.64) yields

$$X = \frac{\dfrac{F_0}{s}}{\sqrt{\left[1 - \left(\dfrac{\omega}{\omega_0}\right)^2\right]^2 + \left(2\zeta \dfrac{\omega}{\omega_0}\right)^2}}. \tag{3.65}$$

Hence, the magnification factor

$$M = \frac{1}{\sqrt{\left[1 - \left(\dfrac{\omega}{\omega_0}\right)^2\right]^2 + \left(2\zeta \dfrac{\omega}{\omega_0}\right)^2}}. \tag{3.66}$$

Similarly, for a damped base-forced mass–spring–dashpot system, the steady-state excursion amplitude is given by

$$X = \frac{b}{\sqrt{\left[1 - \left(\dfrac{\omega}{\omega_0}\right)^2\right]^2 + \left(2\zeta \dfrac{\omega}{\omega_0}\right)^2}}. \tag{3.67}$$

# 3.7   Nonlinear springs

Consider a mass–spring–dashpot system

$$m\ddot{x} + \beta\dot{x} + f(x) = F(t), \tag{3.68}$$

where $f(x)$ is a function representing the variable stiffness of the system,

$$f(x) = s_1 x + s_2 x^3, \tag{3.69}$$

and $F(t)$ is a periodic driving function. The driving function is divided into

$$F(t) = F_1 \cos\omega t + F_2 \sin\omega t, \tag{3.70}$$

so that

$$|F| = \sqrt{F_1^2 + F_2^2}. \tag{3.71}$$

After substituting

$$r = \frac{\beta}{m}, \tag{3.72}$$

$$\epsilon = \frac{s_2}{m}, \tag{3.73}$$

and

$$\omega_0 = \sqrt{\frac{s_1}{m}}, \tag{3.74}$$

(3.68) can be rewritten as

$$\ddot{x} + r\dot{x} + \omega_0^2 x + \epsilon x^3 = \frac{F_1 \cos\omega t + F_2 \sin\omega t}{m}, \tag{3.75}$$

which is referred to as Duffin's equation. Again, we assume the same form as the driving function

$$x = A\cos\omega t. \tag{3.76}$$

Substituting into (3.75) gives

$$\begin{aligned}
\left(\omega_0^2 - \omega^2\right) A\cos\omega t &- A\omega r \sin\omega t + \epsilon A^3 \cos^3\omega t \\
= \left(\omega_0^2 - \omega^2\right) A\cos\omega t &- A\omega r \sin\omega t + \tfrac{3}{4}\epsilon A^3 \cos\omega t + \tfrac{1}{4}\epsilon A^3 \cos 3\omega t \\
= \tfrac{F_1}{m}\cos\omega t &+ \tfrac{F_2}{m}\sin\omega t.
\end{aligned} \tag{3.77}$$

If we neglect the higher harmonic,

$$\left(\omega_0^2 - \omega^2\right) A + \frac{3}{4}\epsilon A^3 = \frac{F_1}{m} \tag{3.78}$$

and

$$-A\omega r = \frac{F_2}{m}. \tag{3.79}$$

Making use of (3.71), we obtain the general form of the relation between oscillation amplitude and driving frequency:

$$\left(\left(\omega_0^2 - \omega^2\right) A + \tfrac{3}{4}\epsilon A^3\right)^2 + A^2\omega^2 r^2 \equiv S(\omega, A)^2 + A^2\omega^2 r^2 = F^2, \tag{3.80}$$

where $S(\omega, A)$ is the so-called response function.

# 4

# Waves and sound
### with Keith Attenborough

## 4.1 Wave equation

To explain the propagation of sound through a medium, we start with a 1-dimensional situation.

Consider an infinitesimal element of length $\mathrm{d}x$ and cross-section $\mathrm{d}S$. Let's assume the element is rigid in all directions except for the $x$-direction. Also, let's define a longitudinal compressive sound wave travelling in the $x$-direction. Suppose that the centre of the element is displaced by a distance $u$ as a result of a sound pressure $p$. Then, the displacements of the boundaries are $\left(u + \frac{\partial u}{\partial x}\frac{\mathrm{d}x}{2}\right)$ and $\left(u - \frac{\partial u}{\partial x}\frac{\mathrm{d}x}{2}\right)$, respectively. The difference in volume, therefore, is

$$\left[\left(u - \frac{\partial u}{\partial x}\frac{\mathrm{d}x}{2}\right) - \left(u + \frac{\partial u}{\partial x}\frac{\mathrm{d}x}{2}\right)\right]\mathrm{d}S = -\frac{\partial u}{\partial x}\,\mathrm{d}x\,\mathrm{d}S. \tag{4.1}$$

The volumetric strain

$$\Delta = \varepsilon_x + \varepsilon_y + \varepsilon_z = -\frac{\partial u}{\partial x}\frac{\mathrm{d}x\,\mathrm{d}S}{\mathrm{d}x\,\mathrm{d}S} = -\frac{\partial u}{\partial x}. \tag{4.2}$$

By definition, the bulk modulus of elasticity $\kappa$ is given by

$$\kappa = \frac{\text{stress}}{\text{strain}} = \frac{p}{\Delta} = -\frac{p}{\partial u/\partial x} \tag{4.3}$$

or

$$p = -\kappa\frac{\partial u}{\partial x}. \tag{4.4}$$

The difference in deformation is caused by a difference in the sound pressure is $p(u - \frac{1}{2}\,dx)$ at one side of the element and $p(u + \frac{1}{2}\,dx)$ at the other side. Hence, the force acting on the element is expressed by

$$- \left[ p(u + \tfrac{1}{2}\,dx) - p(u - \tfrac{1}{2}\,dx) \right] dS = -\frac{\partial p}{\partial x}\,dx\,dS. \tag{4.5}$$

This force accelerates the mass $\rho\,dx\,dS$, with $\rho$ being the density. The acceleration equals the second time derivative of the displacement in the $x$-direction $u$. Therefore, the following balance must hold:

$$-\frac{\partial p}{\partial x}\,dx\,dS = \rho\,\frac{\partial^2 u}{\partial t^2}\,dx\,dS, \tag{4.6}$$

which gives the equation of motion

$$\frac{\partial p}{\partial x} = -\rho\,\frac{\partial^2 u}{\partial t^2}. \tag{4.7}$$

Substituting (4.4) for $p$ into (4.7) results in

$$-\kappa\,\frac{\partial^2 u}{\partial x^2} = -\rho\,\frac{\partial^2 u}{\partial t^2} \tag{4.8}$$

or

$$\frac{\partial^2 u}{\partial t^2} = \frac{\kappa}{\rho}\,\frac{\partial^2 u}{\partial x^2}, \tag{4.9}$$

which is the displacement wave equation of a periodic fluctuation $u$ in the $x$-direction at a speed

$$c = \sqrt{\frac{\kappa}{\rho}}. \tag{4.10}$$

Taking the partial derivative with respect to $x$ of (4.7) gives

$$\frac{\partial^2 p}{\partial x^2} = -\rho\,\frac{\partial^3 u}{\partial x\,\partial t^2}, \tag{4.11}$$

whereas taking the second derivative with respect to $t$ of (4.4) gives

$$\frac{\partial^2 p}{\partial t^2} = -\kappa\,\frac{\partial^3 u}{\partial x\,\partial t^2}. \tag{4.12}$$

Combining (4.11) and (4.12) gives the linear wave equation

$$\frac{\partial^2 p}{\partial t^2} = \frac{\kappa}{\rho}\,\frac{\partial^2 p}{\partial x^2} = c^2\,\frac{\partial^2 p}{\partial x^2}. \tag{4.13}$$

## 4.2  Speed of sound in air

For an adiabatic change in an ideal gas,

$$PV^\gamma = \text{constant}, \tag{4.14}$$

where $P$ is the absolute pressure, $V$ the volume of an element, and $\gamma$ the ratio of specific heats. Differentiating with respect to $V$ gives

$$V^\gamma \frac{\mathrm{d}P}{\mathrm{d}V} + \gamma PV^{\gamma-1} = 0. \tag{4.15}$$

Dividing this through by $V^{\gamma-1}$ gives

$$-\frac{V}{\mathrm{d}V}\,\mathrm{d}P = \gamma P. \tag{4.16}$$

The volumetric strain is

$$\Delta = -\frac{\mathrm{d}V}{V} \tag{4.17}$$

and the stress associated with it is

$$p = \mathrm{d}P. \tag{4.18}$$

Hence, using (4.3), (4.16) can be written as

$$\kappa = \gamma P. \tag{4.19}$$

Therefore, the speed of sound in air

$$c = \sqrt{\frac{\kappa}{\rho}} = \sqrt{\frac{\gamma P}{\rho}}. \tag{4.20}$$

In an ideal gas,

$$PV = n\mathcal{R}\mathcal{T}, \tag{4.21}$$

where $n$ is the amount of substance, $\mathcal{R}$ is the gas constant, and $\mathcal{T}$ is the absolute temperature. Replacing $n$ by $\frac{m}{M}$ and $R$ by $M\bar{\mathcal{R}}$, in which $M$ is the molar mass and $\bar{\mathcal{R}}$ the specific gas constant, we obtain

$$PV = m\bar{\mathcal{R}}\mathcal{T} = \rho V \bar{\mathcal{R}}\mathcal{T}, \tag{4.22}$$

or

$$\frac{P}{\rho} = \bar{\mathcal{R}}\mathcal{T}, \tag{4.23}$$

Thus, for an ideal gas,

$$c = \sqrt{\gamma \bar{\mathcal{R}}\mathcal{T}}. \tag{4.24}$$

## 4.3 Solutions of the 1-dimensional wave equation

Consider the wave equation (4.13)

$$\frac{\partial^2 p}{\partial t^2} = c^2 \frac{\partial^2 p}{\partial x^2} \tag{4.25}$$

and a solution of the form

$$p = f_1(ct - x) + f_2(ct + x). \tag{4.26}$$

Then,

$$\frac{\partial^2 p}{\partial t^2} = c^2 \left( f_1'' + f_2'' \right) \tag{4.27}$$

and

$$\frac{\partial^2 p}{\partial x^2} = f_1'' + f_2''. \tag{4.28}$$

For the displacement equation (4.9), a similar solution is considered:

$$u = g_1(ct - x) + g_2(ct + x). \tag{4.29}$$

If we concentrate on the wave travelling in the positive direction (progressive wave), described by $g_1$, for the moment ignoring $g_2$, at $t = 0$, $u_0 = g_1(-x)$. Also, at $t = 1$, $u_1 = g_1(c - x)$. The displacement $u_1$ caused by $p_1$ must have the same form as the displacement $u_0$ caused by $p_0$, only at a distance $1c$ further on. Hence,

$$g_1(-x_0) = g_1(c - x_1) \tag{4.30}$$

and thus

$$-x_0 = c - x_1 \tag{4.31}$$

or

$$x_1 = x_0 + c. \tag{4.32}$$

For the wave travelling in the negative direction (regressive wave), described by $g_2$, the corresponding equations are

$$g_2(x_0) = g_2(c + x_1) \tag{4.33}$$

and

$$x_1 = x_0 - c. \tag{4.34}$$

Consider a plane single-frequency (monotonous) progressive wave, for which the pressure deviation from the ambient constant value is described by

$$p = p_0 \cos \left[ \omega \left( t - \frac{x}{c} \right) \right] = p_0 \cos \left( \omega t - kx \right). \tag{4.35}$$

Here, $p_0$ is the acoustic pressure amplitude and

$$k = \frac{\omega}{c} \qquad (4.36)$$

is the wave number or propagation constant. Consider the spatial distribution at a fixed time $\omega t = \text{constant}$. Then, if $x$ is equal to the wavelength $\lambda$, $kx$ must equal $2\pi$, i.e.,

$$k = \frac{2\pi}{\lambda}, \qquad (4.37)$$

from which it follows that

$$c = f\lambda. \qquad (4.38)$$

## 4.4   Sound energy

The potential energy in a sound wave is

$$E_\mathrm{P} = -\int p \, \mathrm{d}V. \qquad (4.39)$$

Note the negative sign because a pressure increase causes a volume decrease. Since

$$V = V_0 + \mathrm{d}V = V_0 - V_0 \frac{\mathrm{d}p}{\kappa}, \qquad (4.40)$$

$$\mathrm{d}V = -V_0 \frac{\mathrm{d}p}{\kappa} = -V_0 \frac{\mathrm{d}p}{\rho c^2}. \qquad (4.41)$$

Therefore,

$$E_\mathrm{P} = \int \frac{p V_0}{\rho c^2} \, \mathrm{d}p = \frac{p^2 V_0}{2\rho c^2}. \qquad (4.42)$$

The kinetic energy

$$E_\mathrm{K} = \tfrac{1}{2} \rho V_0 \nu^2. \qquad (4.43)$$

Here, $\nu$ is the particle velocity. To find a relationship between $p$ and $\nu$, we combine equations (4.4) and (4.10):

$$p = -\rho c^2 \frac{\partial u}{\partial x}. \qquad (4.44)$$

Consider (4.35) and (4.36). Since $u$ has the form

$$u = u_0 \cos\left(\omega t - \frac{\omega}{c} x\right), \qquad (4.45)$$

(4.44) can be rewritten as

$$p = -\rho c \omega \, u_0 \sin\left(\omega t - \frac{\omega}{c} x\right) \qquad (4.46)$$

and

$$\nu = \frac{\partial u}{\partial x} = -\omega \, u_0 \sin\left(\omega t - \frac{\omega}{c} x\right). \qquad (4.47)$$

These result in

$$p = \rho\, c\, \nu. \tag{4.48}$$

Therefore,

$$E_{\mathrm{K}} = \frac{p^2 V_0}{2\rho c^2} \tag{4.49}$$

and the total energy

$$E_{\mathrm{T}} = E_{\mathrm{P}} + E_{\mathrm{K}} = \frac{p^2 V_0}{\rho c^2}. \tag{4.50}$$

The energy per unit volume is defined by

$$\mathcal{E} = \frac{E_{\mathrm{T}}}{V} = \frac{p^2}{\rho c^2}. \tag{4.51}$$

The energy that, at any instant, is contained in a column of unit cross sectional area and of length $\mathrm{d}t$ is $\mathcal{E}c\,\mathrm{d}t$. Therefore, the flow of energy, the instantaneous intensity

$$I_t = \mathcal{E}c = \frac{p^2}{\rho c} = p\,\nu. \tag{4.52}$$

The average intensity over a period $T$ is

$$I = \int_0^T p(t)\nu(t)\,\mathrm{d}t = \tfrac{1}{2}p_0\nu_0, \tag{4.53}$$

where $p_0$ is the acoustic pressure amplitude and $\nu_0 = -\omega u_0$ is the amplitude of the particle velocity.

## 4.5  Point and line sources

The sound power of a source is the rate at which the source produces sound energy. It is an intrinsic property. If the power passing through an area $S$ is $W$, the intensity can also be defined as the power per unit area:

$$I = \frac{W}{S}. \tag{4.54}$$

For a point source, the intensity of the sound in point at distance $r$ from the source is

$$I = \frac{W}{4\pi r^2}. \tag{4.55}$$

Thus,

$$I \propto \frac{1}{r^2}. \tag{4.56}$$

Also, quoting (4.52),

$$I \propto p^2. \tag{4.57}$$

Therefore,

$$p \propto \frac{1}{r}. \tag{4.58}$$

The complex representation (*cf.* Section 4.10) of the acoustic pressure is

$$p = \frac{p_0}{r} e^{j(\omega t - kr)}. \tag{4.59}$$

The spherical wave equation is given by

$$\frac{\partial^2}{\partial t^2}(rp) = \frac{\kappa}{\rho} \frac{\partial^2}{\partial r^2}(rp). \tag{4.60}$$

For a cylindrical wave,

$$I = \frac{W}{2\pi r}. \tag{4.61}$$

Hence,

$$p \propto \frac{1}{\sqrt{r}}. \tag{4.62}$$

The acoustic pressure is

$$p = p_0 \, J_0(kr) \, e^{j\omega t}, \tag{4.63}$$

where $J_0$ is the Bessel function of order zero of the first kind, written out in (6.12). The cylindrical wave equation is given by

$$\frac{\partial^2 p}{\partial t^2} = \frac{\kappa}{\rho} \left( \frac{\partial^2 p}{\partial r^2} + \frac{1}{r} \frac{\partial p}{\partial r} \right). \tag{4.64}$$

## 4.6   Doppler effect

If the sound source emitting at frequency $f$ is moving at a velocity $v_s$ towards its audience at rest, the wavelength is reduced to

$$\lambda' = (c - v_s) \, T. \tag{4.65}$$

Hence, the frequency experienced is

$$f' = \frac{f}{1 - \dfrac{v_s}{c}}. \tag{4.66}$$

If the source moves away from the audience, $v_s$ is negative. If the source moves at supersonic speed ($v_s > c$), the wavefront has the shape of a cone with an aperture

$$\sin \frac{\theta}{2} = \frac{c}{v_s}. \tag{4.67}$$

If the audience is moving at a velocity $v_a$ towards the sound source, the frequency experienced is

$$f' = \frac{c + v_a}{\lambda} = \left( 1 + \frac{v_a}{c} \right) f. \tag{4.68}$$

## 4.7 Root-mean-square pressure

The root-mean-square pressure $p_{\text{rms}}$ is defined by

$$p_{\text{rms}}^2 = \frac{1}{T} \int_0^T p^2 \, dt = \frac{1}{T} \int_0^T p_0^2 \cos^2 \omega t \, dt, \tag{4.69}$$

where $T$ is the period. Using $\cos 2\theta = 2\cos^2 \theta - 1$, this simplifies to

$$p_{\text{rms}}^2 = \frac{1}{T} \int_0^T \frac{p_0^2}{2} (\cos 2\omega t + 1) \, dt = \frac{p_0^2}{2T} \left[ t - \frac{\sin 2\omega t}{2\omega} \right]_0^T = \frac{p_0^2}{2}. \tag{4.70}$$

Thus,

$$p_{\text{rms}} = \frac{p_0}{\sqrt{2}} \approx 0.707 p_0 \tag{4.71}$$

for a monotonous wave. In general,

$$\frac{\text{pressure amplitude}}{\text{root-mean-square pressure}} = \frac{p_0}{p_{\text{rms}}} = \text{crest factor.} \tag{4.72}$$

## 4.8 Superposition of waves

Consider, from (4.35) and (4.36), a sound wave consisting of two different frequencies and amplitudes

$$p = p_{0,1} \cos \left[ \omega_1 \left( t - \frac{x}{c} \right) \right] + p_{0,2} \cos \left[ \omega_2 \left( t - \frac{x}{c} \right) \right]. \tag{4.73}$$

Then,

$$\begin{aligned} p^2 &= p_{0,1}^2 \cos^2 \left[ \omega_1 \left( t - \frac{x}{c} \right) \right] + p_{0,2}^2 \cos^2 \left[ \omega_2 \left( t - \frac{x}{c} \right) \right] \\ &\quad + 2 p_{0,1} p_{0,2} \cos \left[ \omega_1 \left( t - \frac{x}{c} \right) \right] \cos \left[ \omega_2 \left( t - \frac{x}{c} \right) \right]. \end{aligned} \tag{4.74}$$

Using $\cos A \cos B = \frac{1}{2} \cos(A + B) + \frac{1}{2} \cos(A - B)$, this becomes

$$\begin{aligned} p^2 &= p_{0,1}^2 \cos^2 \left[ \omega_1 \left( t - \frac{x}{c} \right) \right] + p_{0,2}^2 \cos^2 \left[ \omega_2 \left( t - \frac{x}{c} \right) \right] \\ &\quad + p_{0,1} p_{0,2} \cos \left[ \left( t - \frac{x}{c} \right) (\omega_1 + \omega_2) \right] \\ &\quad + p_{0,1} p_{0,2} \cos \left[ \left( t - \frac{x}{c} \right) (\omega_1 - \omega_2) \right]. \end{aligned} \tag{4.75}$$

Thus,

$$p_{\text{rms}}^2 = \frac{1}{T} \int_0^T p^2 \, dt = p_{1,\text{rms}}^2 + p_{2,\text{rms}}^2. \tag{4.76}$$

This implies that for waves with different frequencies but the same amplitude,

$$p_{rms}^2 = 2p_{1,rms}^2 = 2p_{2,rms}^2. \tag{4.77}$$

For a band of frequencies,

$$p_{rms}^2 = \int p_{f,rms}^2 \, df. \tag{4.78}$$

If the waves have the same frequency, the phase becomes in important. Consider the added waves

$$p = p_{0,1} \cos\left(\omega t - kx + \phi_1\right) + p_{0,2} \cos\left(\omega t - kx + \phi_2\right). \tag{4.79}$$

Then,

$$p_{rms}^2 = p_{1,rms}^2 + p_{2,rms}^2 + 2p_{1,rms}p_{2,rms} \cos\left(\phi_1 - \phi_2\right). \tag{4.80}$$

Obviously, for waves of the same amplitude and phase,

$$p_{rms} = 2p_{1,rms} = 2p_{2,rms}, \tag{4.81}$$

whereas for waves of the same amplitude but opposite phase,

$$p_{rms} = 0. \tag{4.82}$$

## 4.9 Beats

Consider two waves of different frequency and amplitude, now not taking into account any phase difference:

$$p = p_0 \cos \omega_1 t + m \, p_0 \cos \omega_2 t, \tag{4.83}$$

where m is a scalar value. We substitute the average frequency

$$\omega_0 = \tfrac{1}{2}\left(\omega_1 + \omega_2\right), \tag{4.84}$$

substitute the difference frequency

$$\Omega = \omega_1 - \omega_2, \tag{4.85}$$

and make use of $\cos(A + B) = \cos A \cos B - \sin A \sin B$:

$$\begin{aligned} p &= p_0 \cos\left(\omega_0 - \tfrac{1}{2}\Omega\right) t + m \, p_0 \cos\left(\omega_0 + \tfrac{1}{2}\Omega\right) t \\ &= p_0(1 + m) \cos \omega_0 t \cos \tfrac{1}{2}\Omega t + p_0(1 - m) \sin \omega_0 t \sin \tfrac{1}{2}\Omega t. \end{aligned} \tag{4.86}$$

If the amplitudes and frequencies are close to each other, and the difference frequency is in the audible range, beats equal to $\Omega$ are perceived.

## 4.10 Complex representation of a plane, harmonic wave

Both $p_0 \cos(\omega t - kx)$ and $p_0 \cos(\omega t - kx - \frac{1}{2}\pi) = p_0 \sin(\omega t - kx)$ satisfy the wave equation. Since $e^{\frac{j\pi}{2}} = j$, their complex sum also satisfies the wave equation:

$$p_0 \cos(\omega t - kx) + jp_0 \sin(\omega t - kx) = p_0 e^{j(\omega t - kx)}. \tag{4.87}$$

The first time derivative of (4.4) is known as the continuity equation:

$$\frac{\partial p}{\partial t} = -\kappa \frac{\partial u}{\partial x \partial t} = -\kappa \frac{\partial \nu}{\partial x}. \tag{4.88}$$

From (4.87), it is evident that

$$\frac{\partial p}{\partial t} = j\omega p, \tag{4.89}$$

whereas

$$-\frac{\partial \nu}{\partial x} = jk\nu = \frac{j\omega}{c}\nu. \tag{4.90}$$

Combining these gives

$$p = \frac{\kappa}{c}\nu = \frac{\rho c^2}{c}\nu = \rho c\nu, \tag{4.91}$$

which is the same result as (4.48). Hence, the impedance $Z$ can also be expressed in terms of $p$ and $\nu$:

$$Z = \rho c = \frac{p}{\nu}. \tag{4.92}$$

The acoustic impedance is generally expressed in Rayls ($1\,\mathrm{Rayl} = 1\,\mathrm{Pa\,s\,m^{-1}}$), or, more conveniently, in MRayls.

## 4.11 Standing waves

Consider two waves of same amplitude moving in opposite directions:

$$p_1 = p_0 e^{j(\omega t - kx)} \tag{4.93}$$

and

$$p_2 = p_0 e^{j(\omega t + kx)}. \tag{4.94}$$

Then

$$p_1 + p_2 = p_0 e^{j\omega t}\left(e^{-jkx} + e^{jkx}\right) = 2p_0 e^{j\omega t} \cos kx. \tag{4.95}$$

This is the expression for a standing wave with nodes at $x = \frac{1}{2}\left(i + \frac{1}{2}\right)\lambda$ and antinodes at $x = \frac{1}{2}i\lambda$, where $i \in \mathbb{Z}$.

# 4.12 Fourier transform

In acoustics, it is very useful to express a signal $s(t)$ by its frequency content. The (complex) spectral density is obtained by the Fourier transform:

$$S(\omega) = \frac{1}{2\pi} \int_{-\infty}^{\infty} s(t)\, e^{-j\omega t} \, dt. \tag{4.96}$$

A signal can be recovered by the inverse Fourier transform:

$$s(t) = \int_{-\infty}^{\infty} S(\omega)\, e^{j\omega t} \, d\omega. \tag{4.97}$$

# 4.13 Decibel scale

The threshold of normal hearing at 1 kHz $p_h$ is $2 \times 10^{-5}$ Pa. The threshold of pain at 1 kHz is 20 Pa. These extremes represent a hearing range factor of $10^6$ in Pa. The intensity equivalent of the threshold of hearing is

$$I_0 = \frac{p^2}{\rho c} \approx \frac{4 \times 10^{-10}}{4 \times 10^2} = 10^{-12} \ \mathrm{W\,m^{-2}}. \tag{4.98}$$

The power equivalent to the threshold of hearing is taken as $W_0 = 10^{-12}$ W. These thresholds are used as reference quantities when defining sound levels on logarithmic scales:

$$\text{sound power level} \quad \text{SWL} = 10\log_{10} \frac{W}{W_0} \ \text{dB re } W_0;$$

$$\text{intensity level} \quad \text{IL} \ = 10\log_{10} \frac{I}{I_0} \ \text{dB re } I_0;$$

$$\text{sound pressure level} \quad \text{SPL} = 10\log_{10} \frac{p^2}{p_h^2}$$

$$= 20\log_{10} \frac{p}{p_h} \ \text{dB re } p_h. \tag{4.99}$$

For air at room temperature (20 °C) and normal pressure ($1.01 \times 10^5$ Pa), the density $\rho = 1.21$ kg m$^{-3}$ and the sound speed $c = 343$ m s$^{-1}$. Hence, the acoustic impedance

$$Z = \rho c = 415 \ \mathrm{kg\,m^{-2}\,s^{-1}} = 415 \ \text{rayls}. \tag{4.100}$$

Then, the resulting relative intensity

$$\frac{I}{I_0} = \frac{p^2}{Z I_0} = \frac{p^2}{415 \times 10^{-12}} = \left( \frac{p}{2.038 \times 10^{-5}} \right) = \left( \frac{p}{p_h} \right)^2 \left( \frac{1}{1.019} \right)^2, \tag{4.101}$$

so that the intensity level

$$\text{IL} = 10 \log_{10} \frac{I}{I_0} = 10 \log_{10} \left(\frac{p}{p_\text{h}}\right)^2 + 10 \log_{10} \left(\frac{1}{1.019}\right)^2$$

$$= 20 \log_{10} \frac{p}{p_\text{h}} - 20 \log_{10} 1.019 = \text{SPL} - 0.160 \text{ dB}.$$

(4.102)

Hence, the difference between the sound pressure level and the intensity level is negligible.

## 4.13.1 Propagation from a point source

The acoustic power of a point source in a lossless medium is

$$W = 4\pi r^2 I,$$

(4.103)

so that

$$W \propto 4\pi r^2 p^2$$

(4.104)

and

$$W_0 \propto p_\text{h}^2.$$

(4.105)

It follows that the sound power level is related to the sound pressure level:

$$\text{SWL} = 10 \log_{10} \frac{W}{W_0} = 10 \log_{10} \frac{4\pi r^2 p^2}{p_\text{h}^2}$$

$$= 20 \log_{10} \frac{p}{p_\text{h}} + 10 \log_{10} 4\pi r^2 = \text{SPL} + 20 \log_{10} r + 11 \text{ dB}$$

(4.106)

or

$$\text{SPL} = \text{SWL} - 20 \log_{10} r - 11 \text{ dB}.$$

(4.107)

On hard ground,

$$I = \frac{W}{2\pi r^2},$$

(4.108)

resulting in

$$\text{SWL} = \text{SPL} + 20 \log_{10} r + 8 \text{ dB}$$

(4.109)

or

$$\text{SPL} = \text{SWL} - 20 \log_{10} r - 8 \text{ dB}.$$

(4.110)

Similarly, at a junction between a floor and a wall,

$$I = \frac{W}{\pi r^2},$$

(4.111)

resulting in

$$\text{SWL} = \text{SPL} + 20 \log_{10} r + 5 \text{ dB}$$

(4.112)

or

$$\text{SPL} = \text{SWL} - 20 \log_{10} r - 5 \text{ dB}.$$

(4.113)

In general,

$$\text{SPL} = \text{SWL} + 10 \log_{10} \frac{Q}{4\pi r^2},$$

(4.114)

where $Q$ is the directivity factor.

## 4.13.2  Distance doubling

Consider the sound pressure levels $\text{SPL}_1$ at a distance $r_1$ and $\text{SPL}_2$ at a distance $r_2 = 2r_1$ from the source. Then, for a point source,

$$\text{SPL}_1 = \text{SWL} - 20\log_{10} r_1 - 11 \text{ dB} \qquad (4.115)$$

and

$$
\begin{aligned}
\text{SPL}_2 &= \text{SWL} - 20\log_{10} 2r_1 - 11 \text{ dB} \\
&= \text{SWL} - 20\log_{10} 2 - 20\log_{10} r_1 - 11 \text{ dB} \\
&= \text{SPL}_1 - 20\log_{10} 2 \\
&= \text{SPL}_1 - 6 \text{ dB}.
\end{aligned}
\qquad (4.116)
$$

For a line source,

$$\frac{I}{I_0} = \frac{W}{W_0}\frac{1}{2\pi r}. \qquad (4.117)$$

Therefore,

$$
\begin{aligned}
\text{SPL} \approx \text{IL} &= 10\log_{10}\frac{W}{W_0} - 10\log_{10} r - 10\log_{10} 2\pi \\
&= \text{SWL} - 10\log_{10} r - 8 \text{ dB}.
\end{aligned}
\qquad (4.118)
$$

In this case,

$$\text{SPL}_1 = \text{SWL} - 10\log_{10} r_1 - 8 \text{ dB} \qquad (4.119)$$

and

$$
\begin{aligned}
\text{SPL}_2 &= \text{SWL} - 10\log_{10} 2r_1 - 8 \text{ dB} \\
&= \text{SWL} - 10\log_{10} 2 - 10\log_{10} r_1 - 8 \text{ dB} \\
&= \text{SPL}_1 - 10\log_{10} 2 \\
&= \text{SPL}_1 - 3 \text{ dB}.
\end{aligned}
\qquad (4.120)
$$

# 4.14  Vectorial notation for the wave equation

The 3-dimensional wave equation in terms of pressure is given by

$$\ddot{\mathbf{p}} = c^2\,\nabla^2\mathbf{p}. \qquad (4.121)$$

where $c = \sqrt{\dfrac{\text{modulus of elasticity}}{\text{density}}}$ is the phase velocity,

$$
\nabla = \begin{pmatrix} \dfrac{\partial}{\partial x} \\[2mm] \dfrac{\partial}{\partial y} \\[2mm] \dfrac{\partial}{\partial z} \end{pmatrix}, \qquad (4.122)
$$

and

$$\nabla^2 = \nabla \cdot \nabla = \frac{\partial^2}{\partial x^2} + \frac{\partial^2}{\partial y^2} + \frac{\partial^2}{\partial z^2}. \tag{4.123}$$

In vector notation,

$$\mathbf{p} = p_0 e^{j(\omega t - \mathbf{k} \cdot \mathbf{r})}, \tag{4.124}$$

where $\mathbf{k}$ is the wave vector and $\mathbf{r}$ is the distance vector. Note that

$$k = |\mathbf{k}|. \tag{4.125}$$

Rewriting the wave equation in terms of a particle displacement vector $\mathbf{u}$ gives

$$\ddot{\mathbf{u}} = c^2 \nabla^2 \mathbf{u}. \tag{4.126}$$

Here,

$$\mathbf{u} = u_0 e^{j(\omega t - \mathbf{k} \cdot \mathbf{r})}. \tag{4.127}$$

## 4.15   Plane waves in isotropic media

So far, we have dealt with compressive (longitudinal) waves only, which cause displacement parallel to the direction of propagation. Shear (transverse) waves cause displacement perpendicular to the direction of propagation. Since fluids cannot support shear stresses, shear waves can only propagate through solids.

In elastic media, the equation of motion (4.7) is rewritten in terms of the stress tensor:

$$\begin{bmatrix} \sigma_x & \tau_{xy} & \tau_{xz} \\ \tau_{yx} & \sigma_y & \tau_{yz} \\ \tau_{zx} & \tau_{zy} & \sigma_z \end{bmatrix} \begin{bmatrix} \dfrac{\partial}{\partial x} \\ \dfrac{\partial}{\partial y} \\ \dfrac{\partial}{\partial z} \end{bmatrix} = -\rho \begin{bmatrix} \dfrac{\partial^2 u}{\partial t^2} \\ \dfrac{\partial^2 v}{\partial t^2} \\ \dfrac{\partial^2 w}{\partial t^2} \end{bmatrix}. \tag{4.128}$$

Let's restate Hooke's law:

$$[\sigma] = 2G[\varepsilon] + \lambda \Delta [U], \tag{4.129}$$

where $G$ is the shear modulus, $\lambda = \kappa - \frac{2}{3}G$ is Lamé's constant, and

$$[\varepsilon] = \begin{bmatrix} \varepsilon_x & \frac{1}{2}\gamma_{yx} & \frac{1}{2}\gamma_{zx} \\ \frac{1}{2}\gamma_{xy} & \varepsilon_y & \frac{1}{2}\gamma_{zy} \\ \frac{1}{2}\gamma_{xz} & \frac{1}{2}\gamma_{yz} & \varepsilon_z \end{bmatrix}, \tag{4.130}$$

in which the strains and shear strains are given by

$$\varepsilon_x = \frac{\partial u}{\partial x}; \qquad \varepsilon_y = \frac{\partial v}{\partial y}; \qquad \varepsilon_z = \frac{\partial w}{\partial z};$$

$$\gamma_{xy} = \frac{\partial u}{\partial y} + \frac{\partial v}{\partial x}; \quad \gamma_{yz} = \frac{\partial v}{\partial z} + \frac{\partial w}{\partial y}; \quad \gamma_{zx} = \frac{\partial w}{\partial x} + \frac{\partial u}{\partial z}, \tag{4.131}$$

Inserting Hooke's law into (4.128) yields the wave equation for isotropic solids. In vector notation,

$$(\lambda + 2G)\,\nabla(\nabla \cdot \mathbf{u}) - G\,\nabla \times \nabla \times \mathbf{u} = \rho\frac{\partial^2 \mathbf{u}}{\partial t^2}. \tag{4.132}$$

In an isotropic solid, the vector displacement of matter $\mathbf{u}$ can be written as

$$\mathbf{u} = \nabla\phi + \nabla \times \psi. \tag{4.133}$$

The displacement involves a scalar potential $\phi$ and a vector potential $\psi$ resulting from the fact that the movement consists of a translation and a rotation. This means that the wave equation can be split into two equations. One corresponds to the propagation of a compressional (longitudinal) wave, while the other corresponds to a shear (transverse) wave. In terms of the potentials the two equations are:

$$\nabla^2\phi = \frac{1}{c_{\mathrm{p}}^2}\frac{\partial^2\phi}{\partial t^2} \tag{4.134}$$

and

$$\nabla^2\psi = \frac{1}{c_{\mathrm{s}}^2}\frac{\partial^2\psi}{\partial t^2}, \tag{4.135}$$

where $c_{\mathrm{p}}$ and $c_{\mathrm{s}}$ are the phase velocities of the compressional wave and the shear wave, respectively. They are characteristic for the material and are given by

$$c_{\mathrm{p}} = \sqrt{\frac{\lambda + 2G}{\rho}} = \sqrt{\frac{E}{\rho}\frac{1-\nu}{(1+\nu)(1-2\nu)}} \tag{4.136}$$

and

$$c_{\mathrm{s}} = \sqrt{\frac{G}{\rho}} = \sqrt{\frac{E}{\rho}\frac{1}{2(1+\nu)}}, \tag{4.137}$$

where $\nu$ is Poisson's ratio, not the particle velocity. If the material properties are unknown, they can be determined from the compressional and shear wave velocities, using

$$\nu = \frac{\frac{1}{2}\left(\frac{c_{\mathrm{p}}}{c_{\mathrm{s}}}\right)^2 - 1}{\left(\frac{c_{\mathrm{p}}}{c_{\mathrm{s}}}\right)^2 - 1}, \tag{4.138}$$

$$G = \rho\,c_{\mathrm{s}}^2, \tag{4.139}$$

and

$$E = 2(1+\nu)\rho\,c_{\mathrm{s}}^2 = \frac{(1+\nu)(1-2\nu)}{1-\nu}\rho\,c_{\mathrm{p}}^2. \tag{4.140}$$

## 4.16   Waves in fluids

In fluids, no shear deformation can occur. Hence, the shear modulus $G = 0$ and the stress tensor

$$[\sigma] = p[U].$$  (4.141)

Thus, Hooke's law reduces to

$$[\sigma] = \lambda \Delta [U].$$  (4.142)

Furthermore,

$$\kappa = \lambda.$$  (4.143)

Consequently, the speed of sound is given by

$$c_{\mathrm{p}} = \sqrt{\frac{\kappa}{\rho}}.$$  (4.144)

## 4.17   Mechanisms of wave attenuation

All media attenuate sounds, so that the excursion $u$ of a plane wave in the $x$-direction decreases exponentially with the distance as

$$u \propto e^{-\alpha x},$$  (4.145)

where $\alpha$ is the attenuation coefficient. $\alpha$ is usually expressed in $\mathrm{m}^{-1}$ or in Neper/m. Since the acoustic power or the intensity are proportional to the squared amplitude, the corresponding attenuation coefficients become $2\alpha$, as shown in (4.154). The attenuation of a plane wave arises from the scattering of energy from the parallel beam by regular reflection, refraction, and diffraction, and from absorption mechanisms as a result of which the mechanical energy is converted into heat.

Three main mechanisms of sound absorption can be identified:

1. Viscous damping. This corresponds to the friction associated with the relative motion of the particles.

   For a fluid, the viscous damping coefficient is given by

   $$\alpha_{\mathrm{v}} = \frac{\omega^2 \eta}{2\rho c^3},$$  (4.146)

   where $\eta$ is the dynamic viscosity of the fluid. In principle, a second viscosity coefficient can be defined but this is ignored here. Note that $\alpha_{\mathrm{v}} \propto \omega^2$. The dissipation mechanisms can also be accounted for in solids by considering the elastic moduli to be complex.

2. Thermal damping. Here, a fraction of the energy carried by the wave is converted into heat by thermoelasticity. Thus, energy is dissipated in the material by thermal conductivity.

The mechanism of attenuation by thermal conductivity is related to the cycles of compression–dilatation associated with the passage of the wave. These cycles are responsible for the establishment of local thermal gradients within the material. Since the material is thermally conductive, the temperature tends towards uniformity within the material. This phenomenon contributes to the increase of the wave attenuation. The thermal damping $\alpha_\theta \propto \omega^2$, as well. The relative importance of the viscous/thermal attenuation depends on the state of matter. For gases,

$$\frac{\alpha_\theta}{\alpha_v} \approx \frac{3}{8}, \tag{4.147}$$

for liquids,

$$\frac{\alpha_\theta}{\alpha_v} \approx 10^{-3}, \tag{4.148}$$

and for solids,

$$\frac{\alpha_\theta}{\alpha_v} \approx 0. \tag{4.149}$$

This means that the viscous losses in gases are approximately three times more important than the thermal losses, and in some cases thermal effects must be considered. In solids and liquids, the attenuation is mainly due to viscous friction, and thermal effects can be ignored.

3. Molecular relaxation. In this case, the temperature or pressure variations associated with the passage of the wave are responsible for alterations in molecular energy level configurations.

   The cycles of compression–dilatation are also responsible for a modification of the molecular configuration of matter. During the passage of the wave, there may be a transition from one molecular configuration to another. Afterwards, the molecules can revert to their original configuration. This process transfers energy and is responsible for wave attenuation by molecular relaxation. A relaxation time is associated with this phenomenon.

In the linear regime, a solution of the wave equation can be written in complex form as

$$u = u_0 e^{j(\omega t - kx)}. \tag{4.150}$$

where $k$ is the complex wave number including the effect of sound attenuation:

$$k = k_{\text{Re}} - j\alpha. \tag{4.151}$$

Hence,

$$u = u_0 e^{-\alpha x} e^{j(\omega t - k_{\text{Re}} x)} \tag{4.152}$$

and

$$p = p_0 e^{-\alpha x} e^{j(\omega t - k_{\text{Re}} x)} \tag{4.153}$$

It can be shown that the formulae for the characteristic impedance, surface impedance, and reflection and transmission coefficients introduced for real wave numbers later are valid when the complex wave number is used. Since $I \propto p^2$,

$$I = I_0 e^{-2\alpha x}. \tag{4.154}$$

Hence,

$$\begin{aligned}
\text{SPL} \approx \text{IL} = 10\log_{10} e^{-2\alpha x} &= 10 \cdot 0.434\log_e e^{-2\alpha x} \\
&= -8.69\alpha x \text{ dB},
\end{aligned} \tag{4.155}$$

yielding an attenuation $D$ of

$$D = -8.69\alpha \text{ dB m}^{-1}. \tag{4.156}$$

## 4.18 Reflection and transmission

Now that the principles of sound propagation have been explained, we can have a closer look at sound hitting a boundary between two media.

### 4.18.1 Derivation of Snell's law

Huygens' principle, that each point on a wavefront can be treated as a secondary source, is consistent with the period $T$ being constant between media. Since

$$T = \frac{1}{f}, \tag{4.157}$$

the frequency $f$ does not change with a change of media.

Consider oblique incidence of a wavefront on a plane interface, as shown in Figure 4.1. The angle of incidence relative to the normal of the interface is $\theta_i$ and the angle of transmission is $\theta_t$. The sound speed of the incident wavefront is $c_1$ and of the transmitted wavefront $c_2$. We chose such equidistant rays through A, B, C, ..., so that it takes exactly $T$ for the wavefront to advance from (A, B, ...) to (A', B', ...), and again $T$ to advance from (A', B', ...) to (A'', B'', ...). Therefore, the distance AA' = A'A'' = BB' = $c_1 T = \lambda_1$, whereas B'B'' = $c_2 T = \lambda_2$. The shared hypotenuse of the triangles AB'A'' and B'A''B'' is denoted by $d$, so that

$$d\sin\theta_i = c_1 T \tag{4.158}$$

and

$$d\sin\theta_t = c_2 T, \tag{4.159}$$

from which it follows, that

$$\frac{\sin\theta_i}{\sin\theta_t} = \frac{c_1}{c_2} = \frac{\lambda_1}{\lambda_2} = \frac{k_t}{k_i} \tag{4.160}$$

or

$$k_i \sin\theta_i = k_t \sin\theta_t. \tag{4.161}$$

This is known as Snell's law.

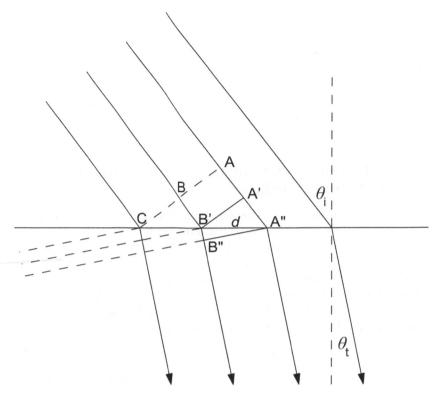

Figure 4.1: Incident rays on a plane interface.

## 4.18.2 Critical angle

In the case $c_2 > c_1$, there exists a critical angle of $\theta_i$ at which transmission occurs parallel to the interface, *i.e.*,

$$\sin \theta_t = 1 \tag{4.162}$$

or

$$\cos \theta_t = 0. \tag{4.163}$$

Since

$$\cos \theta_t = \sqrt{1 - \sin^2 \theta_t} = \sqrt{1 - \left(\frac{c_2}{c_1}\right)^2 \sin^2 \theta_i}, \tag{4.164}$$

the critical incidence angle is found at

$$\theta_i = \arcsin \frac{c_1}{c_2}. \tag{4.165}$$

The index of refraction is defined by

$$n = \frac{c_1}{c_2}. \tag{4.166}$$

For n≫ 1, according to (4.164),

$$\cos \theta_t \approx 1 \qquad (4.167)$$

and thus

$$\theta_t \approx 0. \qquad (4.168)$$

This means that the refracted sound travels in the direction of the normal to the surface, whatever the angle of the incident sound is. The surface is then said to be locally reacting or to satisfy and impedance condition.

### 4.18.3 Reflection and transmission of waves on a plane fluid–fluid interface

Consider oblique incidence on a plane fluid–fluid interface, resulting in reflected and transmitted waves. The fluids are characterised by their respective densities and sound speeds $c_1$ and $c_2$. It is convenient to refer to the products of these quantities, the characteristic or wave impedances $Z_1$ and $Z_2$. These represent the ratios between pressure and velocity at any point in the wave. At a given angle of incidence $\theta_i$ relative to the normal of the interface, the angles of reflection $\theta_r$ and transmission $\theta_t$ are found by writing Snell's law of conservation of the components of the wave numbers along the interface:

$$k_i \sin \theta_i = k_r \sin \theta_r = k_t \sin \theta_t, \qquad (4.169)$$

where the wave numbers of the incident wave $k_i$ and of the reflected wave $k_r$ are

$$k_i = k_r = \frac{\omega}{c_1} \equiv k_1 \qquad (4.170)$$

and the wave number of the transmitted wave is

$$k_t = \frac{\omega}{c_2} \equiv k_2. \qquad (4.171)$$

This implies that

$$\theta_i = \theta_r \equiv \theta_1 \qquad (4.172)$$

and

$$\frac{\sin \theta_1}{c_1} = \frac{\sin \theta_2}{c_2}, \qquad (4.173)$$

where $\theta_2 \equiv \theta_t$.

### 4.18.4 Reflection and transmission coefficients

Two basic continuity conditions must be satisfied at the interface ($x = 0$) at a given time ($t = $ constant):

1. Continuity of velocity: The normal components of the particle velocity must be equal on either side of the interface:

$$\nu_i \cos \theta_1 - \nu_r \cos \theta_1 = \nu_t \cos \theta_2. \qquad (4.174)$$

2. Continuity of pressure: The pressure variations must be equal on either side of the interface:

$$p_i + p_r = p_t, \tag{4.175}$$

where

$$\begin{aligned} p_i &= A_1\, e^{-jk_1 x}; \\ p_r &= B_1\, e^{jk_1 x}; \\ p_t &= A_2\, e^{-jk_2 x}. \end{aligned} \tag{4.176}$$

Since $x = 0$,

$$\begin{aligned} p_i &= A_1; \\ p_r &= B_1; \\ p_t &= A_2. \end{aligned} \tag{4.177}$$

Thus, (4.175) reduces to

$$A_1 + B_1 = A_2. \tag{4.178}$$

Using (4.92), (4.174) becomes

$$\frac{A_1}{Z_1} \cos\theta_1 - \frac{B_1}{Z_1} \cos\theta_1 = \frac{A_2}{Z_2} \cos\theta_2. \tag{4.179}$$

The pressure reflection coefficient is defined by

$$R = \frac{B_1}{A_1} \tag{4.180}$$

and the pressure transmission coefficient by

$$T = \frac{A_2}{A_1}, \tag{4.181}$$

so that

$$T = R + 1. \tag{4.182}$$

Combining (4.178) and (4.179) yields

$$R = \frac{\frac{Z_2}{\cos\theta_2} - \frac{Z_1}{\cos\theta_1}}{\frac{Z_2}{\cos\theta_2} + \frac{Z_1}{\cos\theta_1}} = \frac{Z_2 \cos\theta_1 - Z_1 \cos\theta_2}{Z_2 \cos\theta_1 + Z_1 \cos\theta_2} \tag{4.183}$$

and

$$T = \frac{2\frac{Z_2}{\cos\theta_2}}{\frac{Z_2}{\cos\theta_2} + \frac{Z_1}{\cos\theta_1}} = \frac{2Z_2 \cos\theta_1}{Z_2 \cos\theta_1 + Z_1 \cos\theta_2}. \tag{4.184}$$

The sound intensity reflection coefficient equals the sound power reflection coefficient and is given by

$$R_I = \frac{I_r}{I_i} = \left(\frac{B_1}{A_1}\right)^2 = R^2 = \frac{(Z_2 \cos\theta_1 - Z_1 \cos\theta_2)^2}{(Z_2 \cos\theta_1 + Z_1 \cos\theta_2)^2}, \tag{4.185}$$

where $I_i$ is the incident intensity and $I_r$ is the reflected intensity. Owing to conservation of energy,

$$R_I + T_I = 1, \tag{4.186}$$

where $T_I$ is the sound intensity transmission coefficient, which equals the sound power transmission coefficient,

$$T_I = \frac{I_r}{I_i} = \frac{4Z_2 Z_1 \cos^2 \theta_1}{(Z_2 \cos \theta_1 + Z_1 \cos \theta_2)^2} = \left(\frac{A_2}{A_1}\right)^2 \frac{Z_1}{Z_2}, \tag{4.187}$$

where $I_t$ is the transmitted intensity.

## 4.18.5   Normal incidence

At normal incidence, the cosines are equal to 1 and the expressions for R and T are simplified to

$$R = \frac{Z_2 - Z_1}{Z_2 + Z_1} \tag{4.188}$$

and

$$T = \frac{2Z_2}{Z_2 + Z_1}. \tag{4.189}$$

A good transmission means a low reflection coefficient. The sound intensity reflection coefficient is simplified to

$$R_I = \left(\frac{Z_2 - Z_1}{Z_2 + Z_1}\right)^2, \tag{4.190}$$

whereas the sound intensity transmission coefficient is simplified to

$$T_I = \frac{4Z_2 Z_1}{(Z_2 + Z_1)^2}. \tag{4.191}$$

The energy loss when passing from steel to air is 99.96%. However, the transmission can be improved by interposing a layer of material on the interface. The idea is to reduce the difference of impedance by inserting an intermediate value for the impedance.

## 4.18.6   Normal incidence on a wall (two fluid–fluid boundaries)

Consider normal incidence on a system of two plane interfaces, at $x = 0$ and $x = l$, respectively. At the first interface, the basic continuity conditions are formulated as

$$A_1 + B_1 = A_2 + B_2 \tag{4.192}$$

and

$$\frac{A_1}{Z_1} - \frac{B_1}{Z_1} = \frac{A_2}{Z_2} - \frac{B_2}{Z_2}. \tag{4.193}$$

At the second interface, the basic continuity conditions are formulated as

$$A_2 e^{-jk_2 l} + B_2 e^{jk_2 l} = A_3 e^{-jk_2 l} \tag{4.194}$$

and

$$\frac{A_2 e^{-jk_2 l}}{Z_2} - \frac{B_2 e^{jk_2 l}}{Z_2} = \frac{A_3 e^{-jk_2 l}}{Z_3}. \tag{4.195}$$

This system of equations can be solved to give the following expression for the power transmission coefficient:

$$T_I = \left(\frac{A_3}{A_1}\right)^2 \frac{Z_1}{Z_3} = \frac{4 Z_3 Z_1}{(Z_3 + Z_1)^2 \cos^2 k_2 l + \left(Z_2 + \frac{Z_1 Z_3}{Z_2}\right)^2 \sin^2 k_2 l}. \tag{4.196}$$

### 4.18.7   Impedance of a rigid-backed fluid layer

If the third medium is rigid, the condition of pressure continuity at the second interface is formulated as

$$A_2 e^{-jk_2 l} + B_2 e^{jk_2 l} = 0, \tag{4.197}$$

whereas the condition of velocity continuity does not hold, since $Z_3 = 0$.

$$\frac{A_2 e^{-jk_2 l}}{Z_2} - \frac{B_2 e^{jk_2 l}}{Z_2} = 0. \tag{4.198}$$

Hence,

$$B_2 = -A_2 e^{-2jk_2 l}. \tag{4.199}$$

Substituting this in (4.192) and (4.193) yields

$$A_1 + B_1 = A_2 \left(1 - e^{-2jk_2 l}\right) \tag{4.200}$$

and

$$A_1 - B_1 = \frac{Z_1}{Z_2} A_2 \left(1 + e^{-2jk_2 l}\right), \tag{4.201}$$

respectively. The ratio of these equations is referred to as the relative surface impedance. It represents the ratio of the acoustic pressure to the particle velocity at $x = 0$.

$$Z_s = \frac{A_1 + B_1}{A_1 - B_1} = \frac{Z_2}{Z_1} \frac{1 - e^{-2jk_2 l}}{1 + e^{-2jk_2 l}} = \frac{Z_2}{Z_1} \frac{e^{jk_2 l} - e^{-jk_2 l}}{e^{jk_2 l} + e^{-jk_2 l}} \tag{4.202}$$

Note that

$$\cosh \phi = \frac{e^\phi - e^{-\phi}}{2} \tag{4.203}$$

and

$$\sinh \phi = \frac{e^\phi + e^{-\phi}}{2}. \tag{4.204}$$

Therefore, (4.202) can be reduced to

$$Z_s = \frac{Z_2}{Z_1} \coth j k_2 l = Z_c \coth j k_2 l, \tag{4.205}$$

where $Z_c = \frac{Z_2}{Z_1}$ is the relative characteristic impedance of layer 2. Its inverse, $\beta_c = \frac{1}{Z_c}$ is the relative characteristic admittance.

## 4.19 Scattering

For structures with radii $r$ much less than the acoustic wavelength, such as red blood cells, the ultrasonic backscattering coefficient is

$$\eta(\omega) \propto k^4 r^6 \left( \frac{\kappa_1 - \kappa_0}{\kappa_0} - \frac{\rho_1 - \rho_0}{\rho_0} \right)^2, \tag{4.206}$$

where $k$ is the acoustic wave number, $\kappa_1$ is the compressibility of the scatterer, $\kappa_0$ is the compressibility of the surrounding medium, $\rho_1$ is the density of the scatterer, and $\rho_0$ is the density of the surrounding medium.

Integrating the intensity of the scattered wave over all directions yields the scattered power $P_s$. The acoustic size $Q_s$ of an object is defined as the scattered power divided by the incident wave intensity $I_i$:

$$Q_s = \frac{P_s}{I_i}. \tag{4.207}$$

Since the density and compressibility parameters of blood cells hardly differ from those of plasma, in the diagnostic ultrasonic frequency range, blood cells are poor scatterers. So-called ultrasound contrast agents can be injected that help to differentiate between blood and other tissue types, by providing additional and desirably characteristic backscatter. Gas microbubbles are suitable contrast agents because of their high compressibility and low density compared with the surrounding medium.

## 4.20 Nonlinear propagation

The actual speed of a propagating sound wave is not just determined by the elasticity of the medium, but also by the instantaneous density and the instantaneous particle velocity $\nu$ in the medium. The sound wave changes shape because of these. The phase velocity of a sound wave can be expressed as:

$$c_\nu = c + \left( 1 + \frac{1}{2} \frac{B}{A} \right) \nu, \tag{4.208}$$

where $A$ and $B$ are temperature-dependent quantities. $B/A$ has been generally referred to as the nonlinearity parameter. From this equation, the distance at which the waveform has become a perfect saw-tooth immediately follows:

$$x_\infty = \frac{\pi c^2}{2\nu_0 \omega \left(1 + \frac{B}{2A}\right)},$$

(4.209)

where $\nu_0$ is the amplitude of the particle velocity.

# 5

# Transducers
Andrew Hurrell

Having thoroughly explored the underlying theoretical basis of ultrasonic propagation in previous chapters, it is now necessary to consider how ultrasound could be used in practice. Clearly this requires devices capable of transmitting and receiving ultrasound, and thus it is necessary to consider transduction mechanisms.

Within the field of medical ultrasonics the term "transducer" is often used to refer to a device that has some transmit capability (whether that be transmit-only or transmit/receive functionality). Ultrasonic devices intended to operate in a receive-only mode are usually referred to as hydrophones. This distinction may seem a little arbitrary, but there is a good practical reason behind it. All medical ultrasound devices need to be capable of generating ultrasound. Some devices (particularly therapeutic ones) are exclusively transmitting devices, whereas diagnostic devices need to be able to transmit and receive ultrasound. A receive-only device is only of interest when attempting to quantify the ultrasonic field produced by an external source of ultrasound. Thus to distinguish between a source and a receiver, the terms transducer and hydrophone are used respectively. Furthermore, ultrasonic transducers may have a wide range of users (*e.g.* physiotherapists, ultrasonographers, researchers, medical physicists) whilst hydrophones are likely to have a much more limited user base, specifically those interested in quantifying ultrasonic fields.

The following chapter has been prepared with this distinction in mind, and, whilst some of the topics covered herein may be equally applicable to both hydrophones and transducers, others may not. The primary goal of this chapter is to provide a comprehensive discussion of devices that are used to generate ultrasound.

# 5.1 The piezo-electric effect

## 5.1.1 Overview of piezo-electricity

The vast majority of ultrasonic transducers incorporate a piezo-electric element. All piezo-electric materials contain an asymmetry in their internal structure such that the centre of positive charges is offset relative to the centre of negative charges. If a deformation is applied to such a material there will be a movement of positive charges relative to the negative ones, and a dipole moment proportional to the applied strain will be developed. This phenomena is called the piezo-electric effect. The Greek πιέζω (piezo) means to press. It is this process that allows receiving devices to produce a voltage signal in response to an incoming ultrasonic wave. Similarly if an electric field is applied to the piezo-electric material, the positive and negative charges will exhibit different displacements. Since these displacements are not equal and opposite, the material will change its external dimensions due to the internal strain. This is the inverse piezo-electric effect and it is this mechanism that is exploited when an ultrasonic transducer is used as a transmitter. The behaviour of piezo-electric materials is therefore described by three categories of material constants:

- Mechanical constants, which affect the purely acoustic processes;

- Electrical constants, which affect the purely electric processes;

- Piezo-electro constants, which affect the conversion between mechanical and electrical forms.

## 5.1.2 Piezo-electric nomenclature

Many of the physical quantities used to describe piezo-electric behaviour are common to this and other chapters. However, in order to avoid confusion, it has been necessary to adopt a nomenclature more common in electronics. Table 5.1 summarises the principal variables and constants used in this chapter.

The electro-mechanical transformation that occurs within a piezo-electric material results in behaviour that can at first seem unexpected. If the electrodes of a piezo-electric material are shorted together, there is a means by which current can flow from one electrode to the other in response to an applied stress. In contrast, when the electrodes are left open circuit, this current cannot flow and internal electrostatic forces attempt to resist the motion of charges caused by the applied stress. Thus an open circuit piezo-electric material appears stiffer that it does when its electrodes are shorted together. Similarly the dielectric behaviour of a piezo-electric materials that is free to vibrate will be different from one that is clamped (and thus limited in its ability to vibrate).

For this reason, material constants describing the purely electrical properties of a piezo-electric material have two forms; one for when the material is free, and the other for when it is clamped. Equally, the purely mechanical properties of a piezo-electric material have two forms: one for when the material is open circuit

| Symbol | Unit | Physical Quantity |
|--------|------|-------------------|
| $D$ | $\mathrm{C\,m^{-2}}$ | Electric displacement |
| $E$ | $\mathrm{V\,m^{-1}}$ | Electric field |
| $S$ | | Strain |
| $T$ | $\mathrm{N\,m^{-2}}$ | Stress |
| $Y$ | $\mathrm{N\,m^{-2}}$ | Young's modulus |
| $c$ | $\mathrm{N\,m^{-2}}$ | Elastic stiffness constant |
| $d$ | $\mathrm{C\,N^{-1}}$ or $\mathrm{m\,V^{-1}}$ | Piezo-electric strain/charge constant |
| $e$ | $\mathrm{C\,m^{-2}}$ or $\mathrm{N\,V^{-1}\,m^{-1}}$ | Piezo-electric stress constant |
| $g$ | $\mathrm{V\,m\,N^{-1}}$ or $\mathrm{m^2\,C^{-1}}$ | Piezo-electric voltage constant |
| $k$ | | Electro-mechanical coupling constant |
| $s$ | $\mathrm{m^2\,N^{-1}}$ | Elastic compliance constant |
| $\epsilon$ | $\mathrm{F\,m^{-1}}$ | Dielectric constant |

Table 5.1: Nomenclature used in this chapter to describe piezo-electric materials.

(no applied electric field), and the other for when it is short circuit (no electrical displacement/charge density). A superscripted variable is used to identify the condition at which a materials constant is determined. Table 5.2 indicates the superscripting conventions that are used throughout this chapter.

| Superscript Variable | Physical Quantity | Implied Condition | Example |
|----------------------|-------------------|-------------------|---------|
| T | Stress | $T = 0$ (free) | $\epsilon^{\mathrm{T}}$ |
| S | Strain | $S = 0$ (clamped) | $\epsilon^{\mathrm{S}}$ |
| E | Electric Field | $E = 0$ (open-circuit) | $c_{33}^{\mathrm{E}}$ |
| D | Electrical Displacement | $D = 0$ (short-circuit) | $c_{33}^{\mathrm{D}}$ |

Table 5.2: Nomenclature used in the context of piezo-electric materials.

Hooke's law (discussed in Section 2.9), can be expressed in the alternative nomenclature as

$$[T] = [c][S]. \tag{5.1}$$

A simple expression such as (5.1) can disguise the complexity of Hooke's law, and it is useful to take a moment and consider the situation more carefully. A full expansion of (5.1) to include all of the stress and strain components is

$$
\begin{bmatrix}
T_{xx} \\
T_{xy} \\
T_{xz} \\
T_{yx} \\
T_{yy} \\
T_{yz} \\
T_{zx} \\
T_{zy} \\
T_{zz}
\end{bmatrix}
= [c]
\begin{bmatrix}
S_{xx} \\
S_{xy} \\
S_{xz} \\
S_{yx} \\
S_{yy} \\
S_{yz} \\
S_{zx} \\
S_{zy} \\
S_{zz}
\end{bmatrix},
\tag{5.2}
$$

where $[c]$ is a $9 \times 9$ matrix. As discussed in Section 2.1 (Equation 2.1) many diagonal components are equal. Incorporating these simplifications reduces (5.2) to

$$
\begin{bmatrix}
T_{xx} \\
T_{yy} \\
T_{zz} \\
T_{yz} \\
T_{zx} \\
T_{xy}
\end{bmatrix}
=
\begin{bmatrix}
c_{xxxx} & c_{xxyy} & c_{xxzz} & c_{xxyz} & c_{xxzx} & c_{xxxy} \\
c_{yyxx} & c_{yyyy} & c_{yyzz} & c_{yyyz} & c_{yyzx} & c_{yyxy} \\
c_{zzxx} & c_{zzyy} & c_{zzzz} & c_{zzyz} & c_{zzzx} & c_{zzxy} \\
c_{yzxx} & c_{yzyy} & c_{yzzz} & c_{yzyz} & c_{yzzx} & c_{yzxy} \\
c_{zxxx} & c_{zxyy} & c_{zxzz} & c_{zxyz} & c_{zxzx} & c_{zxxy} \\
c_{xyxx} & c_{xyyy} & c_{xyzz} & c_{xyyz} & c_{xyzx} & c_{xyxy}
\end{bmatrix}
\begin{bmatrix}
S_{xx} \\
S_{yy} \\
S_{zz} \\
S_{yz} \\
S_{zx} \\
S_{xy}
\end{bmatrix}.
\tag{5.3}
$$

The notation used in the discussion above is somewhat cumbersome. To address this problem the reduced notation of ANSI/IEEE 176[1] will be adopted. Table 5.3 summarises the reduced notation.

| Direction Subscript | Numeric Subscript |
| --- | --- |
| $xx$ | 1 |
| $yy$ | 2 |
| $zz$ | 3 |
| $yz$ or $zy$ | 4 |
| $zx$ or $xz$ | 5 |
| $xy$ or $yx$ | 6 |

Table 5.3: Reduced notation.

Kolsky[2] discusses the need for elastic energy to be a single-valued quantity, and shows that this can only happen when there is symmetry in all non-diagonal

[1]ANSI/IEEE 176. *ANSI/IEEE Standard on piezoelectricity*. New York: IEEE **1987**.
[2]Kolsky H. *Stress Waves in Solids*. New York: Dover **1963**.

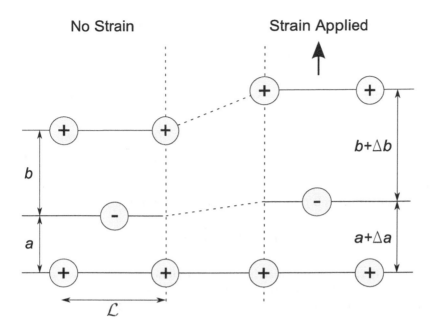

Figure 5.1: How an applied strain affects a piezo-electric material.

elements of the stiffness matrix $[c]$ (*i.e.*, $c_{ij} = c_{ji}$ for $i, j \in 1, 2, 3, 4, 5, 6$ and $i \neq j$). This means that, without loss of generality, $[c]$ now contains a maximum of 21 independent constants. Wherever a structure has further symmetry, additionally simplifications of $[c]$ may be employed. For this reason an isotropic material has only two independent constants (Lamé's constants).

## 5.1.3   Piezo-electric constitutive equations

Having introduced the concept of piezo-electricity the fundamental equations that describe this effect will now be considered. The derivation of these equations follows the method established by Kino.[3] Consider a piezo-electric material as shown in Figure 5.1 with charge of magnitude $+q$ on the positive charges and $-q$ on the negative ones. The asymmetry described earlier means that the unstrained inter-row spacings are $a$ and $b$. The inter-column spacing in both the in-plane and out-of-plane directions is $\mathcal{L}$. In the unstrained state, the polarization ($P$) of any given cell is

$$P = \frac{\text{dipole strength}}{\text{volume}} = \frac{qa - qb}{\mathcal{L}^2(a + b)} = \frac{q(a - b)}{\mathcal{L}^2(a + b)}. \tag{5.4}$$

[3]Kino GS. *Acoustic Waves: Devices, Imaging, and Analog Signal Processing.* Upper Saddle River: Prentice-Hall **1987**.

When a 1-dimensional strain is applied, the polarization becomes

$$P + \Delta P = \frac{q(a + \Delta a) - q(b + \Delta b)}{\mathcal{L}^2(a + \Delta a + b + \Delta b)}. \tag{5.5}$$

To a first order approximation $\Delta a = aS$ where S is the applied stain; a similar approximation is valid for $\Delta b$. Therefore the change in polarization is

$$\Delta P = (P + \Delta P) - P = \frac{q[(a + aS) - (b + bS) - (a - b)]}{\mathcal{L}^2(a + \Delta a + b + \Delta b)}. \tag{5.6}$$

If it is assumed that $\mathcal{L}^2(\Delta a + \Delta b)$ makes a negligible contribution to the volume of the strained unit cell then incorporating (5.4) into (5.6) yields

$$\Delta P = \frac{q(a - b)}{\mathcal{L}^2(a + b)} S = eS. \tag{5.7}$$

This simple derivation was restricted to one dimension; however, as has been seen already, strain is a $3 \times 3$ tensor. The generalisation of (5.7) to cater for full 3-dimensional interactions is

$$[\Delta P] = [e][S]. \tag{5.8}$$

For a dielectric material the electrical displacement field is defined as

$$[D] = [\epsilon][E] + [\Delta P]. \tag{5.9}$$

Incorporating (5.8) into (5.9) at constant strain leads to

$$[D] = [\epsilon^S][E] + [e][S], \tag{5.10}$$

where $e$ is the piezoelectric stress constant.

The force on a charge due to the presence of an electric field is

$$F = q\,E. \tag{5.11}$$

The stress on a positive charge in the upper row is given by

$$T_b = \frac{\text{Force}}{\text{Area}} = \frac{+qE}{\mathcal{L}^2}, \tag{5.12}$$

whilst the stress on a positive charge in the lower row is given by

$$T_a = \frac{-qE}{\mathcal{L}^2}. \tag{5.13}$$

Therefore the average stress due to the presence of an electrical field within the material $T_E$ is given by

$$T_E = \frac{aT_a + bT_b}{a + b} = \frac{\frac{-aqE}{\mathcal{L}^2} + \frac{bqE}{\mathcal{L}^2}}{a + b} = \frac{\frac{q(b-a)}{\mathcal{L}^2}}{b + a} = eE. \tag{5.14}$$

The total stress within the material will be the combination of the externally applied stress, $T$, and the stress due to the electrical field, $T_E$. Generalising to three dimensions and applying Hooke's law (5.1) yields

$$[T] + [T_E] = [c^E][S].$$ (5.15)

Note that a varying electric field would lead to variations in the stress caused by it. To avoid this complication the elastic constant matrix has become $[c^E]$ to indicate that it has been evaluated at constant electric field. Substituting (5.14) into (5.15) results in

$$[T] = [c^E][S] - [e][E].$$ (5.16)

Equations (5.10) and (5.16) are the stress–charge form of the piezo-electric constitutive equations. Other forms of the piezo-electric equations based upon different combinations of the underlying variables (stress, strain, charge and voltage) are available. An example of the inter-relation between two forms is now given.

Hooke's law (5.1) tells us how stress and strain are related via the elastic constant matrix $[c]$ (or its inverse, the compliance matrix $[s] = [c]^{-1}$) and simple rearrangement of this equation yields

$$[s^E] = \frac{\text{Strain}}{\text{Stress}}.$$ (5.17)

Due to the inherent electro-mechanical coupling of a piezo-electric material (as to be discussed later in Section 5.1.5), a changing electric field would result in a change of the stiffness of the material. So as above, the elastic compliance matrix $[s^E]$ is held at constant (probably zero) electric field. The piezo-electric coefficients will be discussed in detail within Section 5.1.4, but for the moment it is necessary borrow the definition of $d$ and $e$ from this section. Multiplying the first definition of (5.22) by (5.17) yields the first definition of (5.21), and therefore

$$[d] = [s^E][e].$$ (5.18)

Substitution of (5.18) into (5.10) and (5.16) yields (5.19) and (5.20), which are the strain–charge forms of the piezo-electric constitutive equations

$$[S] = [s^E][T] + [d][E]$$ (5.19)

and

$$[D] = [d][T] + [\epsilon^T][E].$$ (5.20)

Note that in changing from (5.10) to (5.20) the independent mechanical variable has changed from strain $[S]$ to stress $[T]$. The calculation of electrical displacement $[D]$ must reflect this change, and thus the dielectric coefficient used to multiply the electric field $[E]$ needs to be evaluated at constant stress rather than constant strain. Hence (5.20) uses $[\epsilon^T]$ rather than $[\epsilon^S]$ as was used in (5.10).

### 5.1.4 Piezo-electric coefficients

Before moving on too far from constitutive equations for piezo-electric materials, it is instructive to consider the piezo constants further. The piezo-electric strain coefficient $d$ is defined as

$$[d] = \frac{\text{Strain developed}}{\text{Applied electric field}} = \frac{\text{Electric displacement developed}}{\text{Applied stress}}. \qquad (5.21)$$

Similarly the piezo-electric stress constant $e$ is defined as

$$[e] = \frac{\text{Stress developed}}{\text{Applied electric field}} = \frac{\text{Electric displacement developed}}{\text{Applied strain}}. \qquad (5.22)$$

Finally the piezo-electric voltage coefficient $g$ is defined as

$$[g] = \frac{\text{Strain developed}}{\text{Applied electric displacement}} = \frac{\text{Electric field developed}}{\text{Applied stress}}. \qquad (5.23)$$

The relationship between $[e]$ and $[d]$ has already been identified in (5.18) where Hooke's law was exploited to convert between stress and strain coefficients. A similar transformation can be used to convert between $[d]$ and $[g]$ coefficients. Note from (5.21) and (5.23) that one equation involves electrical fields whereas the other uses electrical displacements. In the absence of a polarization $\Delta P$, (5.9) can be rearranged to give

$$[\epsilon] = \frac{[D]}{[E]} = \frac{\text{Electrical displacement}}{\text{Electrical field}}. \qquad (5.24)$$

Once again, it is necessary to ensure that there are no changes to electrical quantities arising from electro-mechanical coupling, so the condition $T = 0$ is enforced and it becomes necessary to use $\epsilon^{\mathrm{T}}$ to indicate this. This can be simply expressed as

$$[d] = [\epsilon^{\mathrm{T}}][g]. \qquad (5.25)$$

In the most general case, each of the piezo-electric coefficient tensors can be expressed as a $3 \times 6$ matrix. However as has been seen earlier in this chapter, the effect of symmetry within the structure of a material can result in simplifications. The reader is referred to the work of Rosenbaum[4] and ANSI/IEEE 176[1] for further discussion of the classes of symmetry and their effect on the piezo-electric coefficients.

The individual elements within one of the piezo-electric coefficients tensors merit further consideration. For simplicity the following discussion is based upon the piezo-electric strain coefficient $d$, but the concepts apply equally to the other coefficients as well. Consider a slab of piezo-electric material as shown in Figure 5.2. The upper and lower surfaces of the slab have had thin electrode layers applied to them. These electrodes can then be used to apply an electric

---

[4]Rosenbaum JF. *Bulk Acoustic Wave: Theory and Devices*. Norwood: Artech House **1988**.

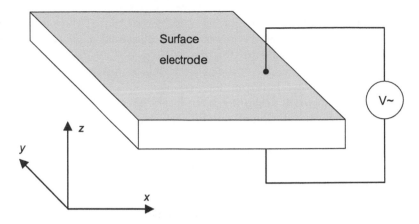

Figure 5.2: Coordinate system for a piezo-electric material.

field across the piezo-electric sample in order to generate a strain, or they can be used to harvest charge that has built up as a result of an externally applied stress. In either case, an electric field in the $z$-direction will be present.

Note that in this simple configuration, there are no electrodes on the other 4 faces of the sample. This means there is no means of making a connection to allow a current to flow in the $x$- and $y$-directions. This means that $D_1$ and $D_2$ must be zero. Recall the second of the strain–charge piezo-electric equations (5.20) and consider specifically the first term $[d][T]$, which deals with the electro-mechanical conversion. When written in full with condensed notation, this term is

$$[d][T] = \begin{bmatrix} d_{11} & d_{12} & d_{13} & d_{14} & d_{15} & d_{16} \\ d_{21} & d_{22} & d_{23} & d_{24} & d_{25} & d_{26} \\ d_{31} & d_{32} & d_{33} & d_{34} & d_{35} & d_{36} \end{bmatrix} \begin{bmatrix} T_1 \\ T_2 \\ T_3 \\ T_4 \\ T_5 \\ T_6 \end{bmatrix}. \qquad (5.26)$$

If a plane stress is applied in the $z$-direction, the component $T_3$ will be non-zero. Since $D_1$ and $D_2$ must be zero, all contributions arising from the first and second rows of the $[d]$ matrix must have a zero sum. Thus it is only the coefficient $d_{33}$ that will determine what electrical displacement is developed in response to this stress. Note that both the applied stress and the electrical response are purely in the $z$-direction; this is precisely what happens when a thin piezo-electric plate is operated as a thickness expander (see Section 5.4.1).

Now consider a plane stress applied in the $x$-direction. In this case the component $T_1$ will be non-zero, and the coefficient $d_{31}$ determines the electrical displacement. Similar arguments apply to $d_{32}$ when a plane stress in the $y$-direction is applied. For this reason the coefficients $d_{31}$ and $d_{32}$ are of primary importance when designing a transducer that will operate as a length expander.

Clearly combinations of these effects are possible, so that if the piezo-electric sample was subject to hydrostatic pressure, the electrical displacement developed in response to this would involve $d_{31} \cdot T_1 + d_{32} \cdot T_2 + d_{33} \cdot T_3$. For these reasons, the coefficients $d_{31}$, $d_{32}$ and $d_{33}$ are widely reported by manufacturers of piezo-electric materials. For completeness, it should be noted that $d_{15}$ and $d_{24}$ are of importance when designing transducers operating in a shear mode.

As has already been mentioned, the $[d]$ coefficients are widely reported and (5.25) can be used to derive the $[g]$ coefficients if required. However, the question "why have multiple piezo-coefficients?" needs addressing. Consider the definitions (5.21) and (5.23); from (5.21) we see that that piezo-electric strain coefficients ($[d]$) tell us what strain (and hence surface displacement) can be expected for an applied electric field. This is precisely the data that would be of use when designing a device to transmit ultrasound. Conversely an ultrasonic receiver needs to maximise the electric field (and hence voltage) developed across the electrodes in response to the stress applied by an incident pressure wave. Therefore the piezo-electric voltage coefficients ($[g]$) are of most use when designing receivers. When designing a device that will both transmit and receive ultrasound, both sets of coefficients will have to be considered.

### 5.1.5 Electro-mechanical coupling coefficient

In the previous section, the definitions of the piezo-electric coefficients expressed the relationships between pairs of variables used in the constitutive equations. During these derivations, care was taken to ensure that the coupling between electrical and mechanical effects was handled appropriately. It is therefore desirable to have one constant that expresses the full interaction between the mechanical, electrical and piezo-electric properties of a material in a simple and concise manner.

Consider a piezo-electric material under open circuit conditions. No current flows from one electrode surface to the other and thus (5.20) becomes

$$[E] = -\frac{[d][T]}{[\epsilon^{\mathrm{T}}]}. \tag{5.27}$$

Substituting (5.27) into (5.19) gives

$$[S] = [s^{\mathrm{E}}][T] + [d]\left(\frac{-[d][T]}{[\epsilon^{\mathrm{T}}]}\right) \tag{5.28}$$

$$= [s^{\mathrm{E}}]\left(1 - \frac{[d]^2}{[s^{\mathrm{E}}][\epsilon^{\mathrm{T}}]}\right)[T]. \tag{5.29}$$

Now, recall Hooke's law (5.1) but recast it in terms of the compliance matrix $[s]$ rather than the stiffness matrix $[c]$ (q.v. (5.17)). This derivation began with the explicit assumption that the piezo-electric material was subject to open circuit conditions. Selecting the stiffness matrix appropriate for this boundary condition gives

$$[S] = [s^{\mathrm{D}}][T]. \tag{5.30}$$

Comparison of (5.30) and (5.29) reveals the relationship between the compliance constants under short-circuit and open-circuit conditions, namely

$$[s^D] = [s^E]\left(1 - \frac{[d]^2}{[s^E][\epsilon^T]}\right) \tag{5.31}$$

$$= [s^E](1 - k^2), \tag{5.32}$$

where the electro-mechanical coupling coefficient $k$ is defined by

$$k^2 = \frac{[d]^2}{[s^E][\epsilon^T]}. \tag{5.33}$$

It is trivial to rearrange (5.31) in terms of elastic constants so that

$$[c^E] = [c^D](1 - k^2). \tag{5.34}$$

This is a highly significant result. Given that $k^2 \geq 0$, the components of $[c^E]$ will always be smaller than their counterparts in $[c^D]$. To understand this, recall that if current cannot flow between the electrodes of a piezo-electric material, the internal electrical stresses make the material appear stiffer than when a current is free to flow. For this reason the matrix $[c^D]$ is often referred to as the "stiffened" elastic constant matrix.

Consider now a clamped piezo-electric material where $[S] = 0$. In this case (5.19) can be rearranged to give

$$[T] = -\frac{[d][E]}{[s^E]}. \tag{5.35}$$

Substituting (5.35) into (5.20) gives

$$[D] = [\epsilon^T][E] - [d]\left(\frac{-[d][E]}{[s^E]}\right) \tag{5.36}$$

$$= [\epsilon^T][E]\left(1 - \frac{[d]^2}{[s^E][\epsilon^T]}\right) \tag{5.37}$$

$$= [\epsilon^T][E]\left(1 - \frac{[d]^2}{[s^E][\epsilon^T]}\right). \tag{5.38}$$

As before, this result needs to be compared with the conditions imposed at the beginning of the derivation. Under clamped conditions, the constant $[\epsilon^S]$ would normally be expected to relate the electric displacement current $[D]$ to the electric field $[E]$. This enables the relationship between the dielectric constants under clamped and free conditions to be defined as

$$[\epsilon^S] = [\epsilon^T]\left(1 - \frac{[d]^2}{[s^E][\epsilon^T]}\right) \tag{5.39}$$

$$= [\epsilon^T](1 - k^2). \tag{5.40}$$

Once again the electro-mechanical coupling coefficient provides the relationship between the two forms of dielectric constant, just as it did with the elastic constants

Before finishing the discussion on electro-mechanical coupling coefficient it is important to remember that the definition of $k^2$ (5.33) is based upon tensor quantities and thus $k^2$ is also a tensor. As was discussed with the piezo-electric coefficients, some of the components of the matrix have particular significance. The designer of a transducer operating in thickness mode will probably be looking to find a material with a high $d_{33}$ value. For this application, the $k_{33}$ component is likely to be the most relevant component of the electro-mechanical coupling matrix. However, the determination of $k_{33}$ imposes no restriction on how the piezo-electric element is moving, so there may also be lateral/radial extension occurring as well. For this reason many piezo-electric component manufacturers also quote $k_t$, which is the purely thickness mode coupling coefficient when the material is laterally clamped.

## 5.2   Piezo-electric materials

### 5.2.1   Piezo-electric ceramic

Piezo-electricity was discovered in 1880 by the Curie brothers. This field has now developed to a stage whereby almost all readers of this text will have a piezo-electric device within a few centimetres of themselves (in the form of a timing circuit in a wrist watch or mobile/cellular phone). Within medical ultrasonics, piezo-ceramics are widely used, although some high frequency applications also exploit piezo-polymer technologies.

There are many commercially available piezo-ceramics such as lead titanate (PT), bismuth titanate, barium titanate, lead metaniobate (PMN), lithium niobate and lead zirconate titanate (PZT). Subtle changes to the constituents and their processing can lead to quite a variation in properties. By way of an example PZT is commonly available in at least five different forms, some of which are characterised as "hard" PZT whilst the others are "soft" PZT. Generally speaking, softer materials have higher levels of internal damping. This leads to the broader-bandwidth, shorter-pulse behaviour (as will be discussed in Section 5.3). In contrast, the harder ceramics have little if any damping and are thus more resonant.

The majority of piezo-electric ceramics exhibit ferroelectric behaviour. A ferroelectric material is one that can have its polarization reversed in response to an external field. Typically the material needs to be heated beyond a critical temperature (known as the Curie temperature) before the dipoles are free to move. Materials of this type are internally organised into a series of domains. Within each domain the internal dipoles are aligned, but there is no net polarisation due to the random orientation of domains with respect to each other. In order to produce a material with a strong piezo-electric behaviour, these domains need to oriented with their polarisations aligned; this is process is referred to as "poling".

A typical poling sequence involves heating the ferro-electric material above the Curie point so that the dipoles become mobile. If an external electric field is applied, the dipoles within the domains will attempt to align themselves with the external field. By then cooling the material to a temperature below the Curie point, the orientation of the polarisations become "frozen" in place. Since all the internal polarisations are now aligned, the material will exhibit a strong piezo-electric response. Critically though, if the material is subsequently heated back above the Curie temperature, the dipoles will become mobile again. If allowed to cool without an external polarising field, the orientations of the dipoles will return to their random state. This means that many piezo-electric materials can be subject to thermal depolarisation when exposed to high temperatures.

| | Units | "Hard" PZT | "Soft" PZT | PVDF | PMN | PT |
|---|---|---|---|---|---|---|
| $d_{33}$ | $\mathrm{pC\,N^{-1}}$ | 328 | 425 | $-33$ | 83 | 70 |
| $d_{31}$ | $\mathrm{pC\,N^{-1}}$ | $-128$ | $-170$ | 23 | $-16$ | $-5.33$ |
| $g_{33}$ | $\mathrm{V\,m\,N^{-1}}$ | 0.028 | 0.0267 | -0.31 | 0.032 | 0.037 |
| $g_{31}$ | $\mathrm{V\,m\,N^{-1}}$ | $-0.011$ | $-0.011$ | 0.210 | $-0.007$ | $-0.003$ |
| $\epsilon_r^T$ | | 1300 | 1800 | 12 | 285 | 215 |
| $c_{33}^D$ | GPa | 158 | 144 | 9.01 | 57.7 | 129 |
| $Y_{33}^D$ | GPa | 95.6 | 84.3 | 0.9 | 92.3 | 163 |
| $k_t$ | | 0.47 | 0.47 | 0.14 | 0.34 | 0.515 |
| $k_{33}$ | | 0.684 | 0.699 | 0.1 | 0.50 | 0.55 |
| $\rho$ | $\mathrm{kg\,m^3}$ | 7700 | 7700 | 1780 | 6200 | 7050 |
| c | $\mathrm{m\,s^{-1}}$ | 4530 | 4324 | 2250 | 3050 | 4270 |
| Z | MRayls | 34.9 | 33.3 | 4.01 | 18.9 | 30.1 |
| $Q_m$ | | $> 1000$ | 70–80 | $< 2$ | $< 15$ | $> 1500$ |

Table 5.4: Typical properties of various piezo-electric materials. Data provided courtesy of: Meggitt SA, Kvistgård, Denmark; MSI, Hampton, VA, USA; Piezo Technologies, Indianapolis, IN, USA.

Most piezo-ceramics have good (or excellent) electrical-to-mechanical conversion capabilities, as can be seen from the data in the first two rows of Table 5.4. Therefore all *should* make good sources of ultrasound. However, having a high [d] coefficient is not the only criteria to consider. Due to both its lower density and lower elastic constants (and hence lower acoustic velocity), PMN has an acoustic impedance that is much smaller than many other piezo-ceramics. This means that it will be easier to transfer energy from PMN into a lower acoustic impedance load material (*e.g.*, water). For the higher acoustic impedance materials (*e.g.*, the PZTs) specific measures are needed (*cf.* Section 5.4.3) to ensure the generated acoustic energy propagates into a water like medium.

The amount of internal damping also needs careful consideration. This can be seen in the value of mechanical Q, $Q_m$, which for PMN is typically less than 15, yet is often more than 1000 for "hard" PZT and PT. A low $Q_m$ is indicative of a material that is is well suited to producing shorter acoustic signals with relatively broad bandwidth. However, a high $Q_m$ should not automatically be seen as a disadvantage. In fact, for those devices operating in a continuous wave mode, low damping is a distinct advantage since is reduces the amount of energy dissipated internally within the ceramic. For this reason therapeutic ultrasound devices, particularly high intensity focussed ultrasound (HIFU) devices, are almost exclusively based on "hard" PZT piezo-ceramics. There are many other factors that influence the particular choice of ceramic for a given application, such as thermal stability, ageing rate, piezo-electric efficiency, and dielectric constant to name but a few.

## 5.2.2    Piezo-electric polymers

Piezo-polymer devices commonly use the polymer polyvinylidene fluoride (PVDF) or its co-polymer with trifluoroethylene (P[VDF-TrFE]). However, some polyvinyl chlorides and co-polymers of Nylon also show piezo-electric activity. Compared with piezo-ceramics, these materials have poor electrical-to-mechanical conversion properties and thus poorer transmit efficiency. Table 5.4 shows the [d] coefficients for PVDF to be an order of magnitude lower than those of the PZTs listed. However piezo-polymers have a much higher receive efficiency (see the [g] coefficients for PVDF). Piezo-polymers also have an acoustic impedance that is much lower than all piezo-ceramics. This means that ultrasonic energy propagating in a low impedance medium (such as water or biological tissue) will find it easier to propagate into a piezo-polymer than it would into a piezo-ceramic. Piezo-polymers are also the most highly damped of all the commonly used piezo-electric materials. These three factors, low $Q_m$, high [g], and low $Z$, mean that piezo-polymers are the ideal choice for broadband, receive-only devices such as hydrophones.

Piezo-polymers are available as very thin films. As will be seen in Section 5.4.1, the thickness of the piezo element is critical in determining the operational frequency of a transducer. High-frequency operation, coupled with the high levels of internal damping, mean that a well-designed piezo-polymer transducer is capable of producing very short acoustic pulses. Thus, despite their poor transmit efficiency, piezo-polymer transducers are commonly used in pulse/echo diagnostic ultrasound systems requiring very high resolution but over acoustic path lengths (*e.g.*, ophthalmic and dermal scanning applications).

Like piezo-ceramics, piezo-polymers need to undergo a poling process. However, many piezo-polymers require an additional orientation process first. To recap, piezo-ceramics contain highly oriented domains and the poling process is needed to ensure consistent alignment of the dipoles within these domains. Piezo-polymers, however, are often semi-amorphous polymers; they contain highly oriented crystallites within a randomly oriented amorphous phase. For these materials, a remnant polarisation can only be imparted if the random orientation polymer chains are "straightened out" prior to poling, and this is commonly done by stretching the material. PVDF is typical of a piezo-polymer that needs to undergo an orientation process. However, some piezo-polymers (such as P[VDF-TrFE] co-polymer) have a inherent dipole moment and do not need to undergo an orientation phase

For those materials that require orientation, stretch ratios can be large — sometimes as much as 5 times the original length. Given that most piezo-polmers are produced as sheets or rolls of film, two possibilities exist for the orientation process. In the first case the film is stretched in one direction only. With relation to the reduced notation of Table 5.3, the thickness of film is the 3-direction, whilst the direction of stretching is the 1-direction. This type of film is referred to as uni-axial film and is characterised by having a large $d_{31}$ coefficient but a small $d_{32}$ coefficient.

Alternatively the film can be stretched equally in the two transverse (*i.e.*, non-thickness) directions. Film processed in this manner is referred to as bi-axial film and has $d_{31} \approx d_{32}$. It should be noted that $d_{31}$ for uni-axial film is approximately 3 times greater than it is for bi-axial film. Interestingly, the $d_{33}$ coefficient appears reasonably constant regardless of which orientation method is used. Piezo-polymers can be poled in a similar method to piezo-ceramics (*i.e.*, heat, apply $E$-field then cool without removing field), but alternative poling methods such as corona poling and hysteresis poling are also used. The reader should consult Sessler,[5] Wang *et al.*,[6] and references therein for further information about piezo-polymers and their preparation.

## 5.3   Transducer bandwidth

All ultrasonic transducers are inherently resonant devices, and therefore can have their output altered by the effect of damping. An undamped resonant system will be capable of producing a large output signal, but only over a narrow frequency range. In contrast, a highly damped system will produce a lower output signal, but will be capable of doing so over a much wider range of frequencies. Quantification of the frequency response of a transducer typically utilises the concept of bandwidth. To calculate bandwidth it is first necessary to determine the peak of the frequency response, and also a threshold level at a fixed reduction (typically $-3\,\mathrm{dB}$ or $-6\,\mathrm{dB}$) relative to the peak. The points at which the frequency response curve crosses this threshold are the lower ($f_\mathrm{l}$) and upper ($f_\mathrm{u}$) cut-off frequencies, and bandwidth ($BW$) is defined as the difference

[5]Sessler G. Piezoelectricity in polyvinylidenefluoride. *J Acoust Soc Am* **1981** 70:1596–1608.

[6]Wang TT, Herbert JM, Glass AM, Eds. *The applications of ferroelectric polymers.* Bishopbriggs: Blackie **1988**.

between these two:

$$BW = f_\mathrm{u} - f_\mathrm{l}. \tag{5.41}$$

The centre frequency ($f_\mathrm{c}$) and fractional bandwidth ($FBW$) are defined as

$$f_\mathrm{c} = \frac{f_\mathrm{u} + f_\mathrm{l}}{2} \text{ and } FBW = \frac{BW}{f_\mathrm{c}} \times 100\%, \tag{5.42}$$

respectively. $FBW$ can be expressed as either a decimal or percentage.

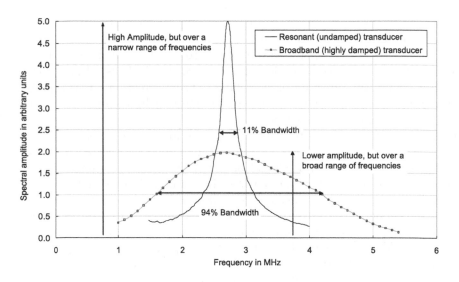

Figure 5.3: The effects of damping on a transducer's frequency response.

Figure 5.3 shows two different transducer responses, to illustrate these concepts. Both response curves in Figure 5.3 have the same centre frequency; however, the peak amplitude of the undamped transducer is much higher than that of its damped counterpart.

The temporal response of the same two transducers is shown in Figure 5.4. The greater signal amplitude of the undamped transducer is clearly visible, but it has a much longer pulse duration. In an imaging application (where short pulses give rise to good axial resolution) the highly damped waveform would be preferable. However, in a therapeutic application (where the aim is the most efficient deposition of ultrasonic energy) the resonant transducer would be the preferred choice.

## 5.4 Transducer construction

When embarking on the design of an ultrasonic transducer, it is useful for the designer to have a simple goal in mind. One such example is:

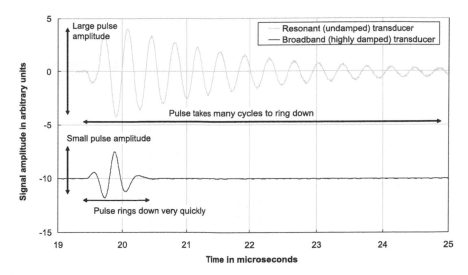

Figure 5.4: The effects of damping on a transducer's frequency response.

> The aim of good transducer design is to maximise the acoustic output from the device, whilst achieving the desired bandwidth and temporal response.

There is no single definition of what is "desired" since that may vary from one device to the next. As previously, a highly resonant transducer is often concerned with maximising output at the expense pulse duration, whereas a transducer in an imaging application will usually require a temporally localised signal to maximise resolution. For this reason, it is often necessary to make compromise decisions during the transducer design process. The remainder of this section is dedicated to a discussion of the various components that go to make up an ultrasonic transducer. A cross-section through a typical single element transducer can be found in Figure 5.5. During the subsequent discussions, important issues relating to each component will be highlighted. This is intended to provide the reader with the ability to make a more informed decision when working towards the design goal.

## 5.4.1 Piezo-electric element

The range of materials available for use as the piezo-active component of a transducer has been discussed in Section 5.2. However, the dimensions of the piezo-electric element also need careful consideration, since they affect the type of resonance exhibited by the transducer, as well as the natural frequency of oscillation. Detailed discussion on the effect of piezo-electric element geometry in relation to resonance modes is provided by Onoe and Tiersten.[7]

---

[7]Onoe M, Tiersten HF. Resonant frequencies of finite piezoelectric ceramic vibrators with high electromechanical coupling. *IEEE Trans Ultrason Eng* **1963** 10:32–38.

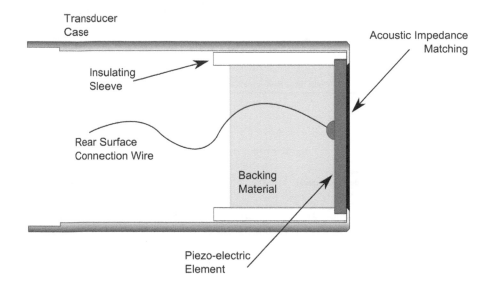

Figure 5.5: Cross section through a typical single element transducer.

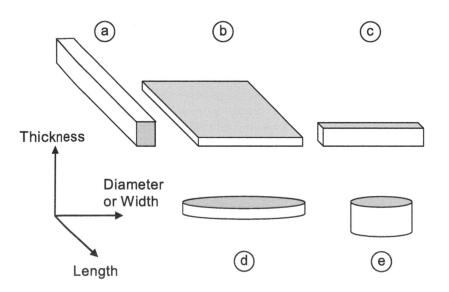

Figure 5.6: A selection of common piezo-electric element shapes. The grey surface indicates the position of the electrode layer.

A selection of the common piezo-electric element shapes can be found in Figure 5.6. Before considering each of these shapes in detail, it is useful to define the aspect ratio of the crystal. This quantity is often used as a convenient way to identify the type of resonance exhibited by a given piezo-electric crystal:

$$\text{Aspect ratio} = \frac{\text{Width or Diameter}}{\text{Thickness}}. \tag{5.43}$$

When a piezo-electric element resonates, its dimensions (in conjunction with various material parameters) will determine what resonance mode is excited; several common resonance modes are discussed later in this chapter. A transducer that excites only one resonance mode will be optimally efficient at the resonant frequency of that mode. If multiple modes are excited, then each will have a different resonant frequency. This may lead to a device with a broader spectrum of operation, although the acoustic output in any one mode will be restricted in comparison to a single mode transducer.

Intuitively it should be possible to derive a single expression relating the resonant frequency of a piezo-electric element to its elastic constants $[c]$, density $\rho$, and dimensions. However, the possibility of multiple resonance modes, and coupling of energy between them, means that this relationship often involves the solution of transcendental equations.[7] For this reason, many piezo-ceramic manufacturers quote frequency constants $N$, so that piezo-electric element dimensions may be appropriately specified to achieve the desired resonant frequency. Common used frequency constants are shown in Table 5.5 along with the resonance mode for which they are of most use.

| Constant | Resonance Mode |
|----------|----------------|
| $N_t$ | TE |
| $N_p$ | RE |
| $N_3$ or $N_{33}$ | LE |
| $N_1$ or $N_{31}$ | WE |

Table 5.5: Examples of commonly specified frequency constants.

All frequency constants are quoted on the assumption that the piezo-electric element is acting as a half-wave resonator. However, as will be discussed in Section 5.4.2, a backing impedance greater than that of the piezo-electric element will force the piezo-electric to behave as a quarter-wave resonant device. In this case the appropriate factor of 2 must be used in conjunction with the frequency constants $N$, to obtain the correct operating frequency.

**Thickness expander (TE)**

When a piezo-electric element has a large lateral extent relative to its thickness, the principle resonance is in the thickness direction. Examples of this can be seen in Figure 5.6 (b) and (d). A resonance of this kind is referred to as a thickness expander (TE). When the aspect ratio falls below 10, other parasitic resonance modes (*e.g.*, radial, circumferential) are excited. The centre frequency and dimensional requirements for operation as a TE are

$$f_c = \frac{N_t}{\text{thickness}} \quad ; \quad \text{aspect ratio} > 10.$$

Therapeutic transducers (*e.g.*, physiotherapy and HIFU) often operate in a TE mode. The vast majority of non-destructive testing (NDT) transducers also operate in this manner.

### Radial expander (RE)

The requirements for operation in RE mode are much the same as for TE, and thus Figures 5.6 (b) and (d) are equally relevant. The applied electric field for RE is in the same direction as TE (*i.e.*, in the thickness direction). However, the resultant extension is from the centre of the plate towards the edge in the radial or width direction. The centre frequency and dimensional requirements for operation as a RE are

$$f_c = \frac{N_p}{\text{thickness}} \quad ; \quad \text{aspect ratio} > 10.$$

### Length expander (LE)

Length expander (LE) resonances occur when a long thin rod has electrodes on its end faces. Both the applied electric field and the resulting extension are in the length direction. An example of a typical LE mode device can be found in Figure 5.6 (a). The centre frequency and dimensional requirements for operation as a LE are

$$f_c = \frac{N_{33}}{\text{length}} \quad ; \quad \text{length} > 2.5 \times \text{ (all other dimensions).}$$

LE resonance is commonly found in piezo-composite transducers, which often involve rods of piezo-ceramic material embedded within a polymeric matrix.

### Width expander (WE)

Width expander (WE) (sometimes called beam expander) resonances typically occur when a flat strip has electrodes across the thickness direction. Figure 5.6 (c) shows a typical WE element. The centre frequency and dimensional requirements for operation as a WE are

$$f_c = \frac{N_{31}}{\text{length}} \quad ; \quad \text{aspect ratio} > 12 \text{ AND width} > 4 \times \text{length.}$$

WE modes are some of the most widely used transducer forms since, in most diagnostic imaging arrays, each element operates as a WE.

## 5.4.2 Backing material

The choice of material used to provide the mechanical termination within an ultrasonic transducer can have a critical effect on the performance of a device. Kossoff[8] was one of the first researchers to provide a comprehensive analysis of this aspect of transducer performance. When an electrical signal is applied to a piezo-electric element, conservation of momentum requires that the crystal move symmetrically about its centreline, and therefore two acoustic signals are generated: one travelling towards the transducer front-face and the other propagating towards the transducer backing. Unless the backing material has the same acoustic impedance as the piezo-electric material, there will be a reflection of the backward travelling signal at the interface with the backing. Consider a backing material with acoustic impedance lower than that of the piezo-electric element. The phase change ($\phi$) of a wave, with wavelength $\lambda$, due to transit across a material of thickness $l$ is given by

$$\phi = \frac{2\pi l}{\lambda}. \tag{5.44}$$

The forward travelling wave experiences a phase change $\phi_F$ as it travels from the centreline to the front face of the piezo element given by

$$\phi_F = \frac{1}{2}\frac{2\pi l}{\lambda} = \frac{\pi l}{\lambda}. \tag{5.45}$$

The rear travelling wave experiences a similar phase change as it propagates. Recall from (4.188) that the pressure reflection coefficient at normal incidence, R, is given by

$$R = \frac{Z_2 - Z_1}{Z_2 + Z_1}, \tag{5.46}$$

where $Z_1$ is the medium the wave is travelling from and $Z_2$ is the medium that the wave is travelling into. From (5.46) it can be seen that R will be negative when going from a medium of high acoustic impedance to a one of lower impedance (*i.e.*, $Z_1 > Z_2$). This is indicative of a phase change of $\pi$ radians upon reflection. The wave also experiences a further phase change during propagation towards the front face of the piezo-electric element. The total phase change experienced by this wave ($\phi_B$) is

$$\phi_B = \frac{\pi l}{\lambda} + \pi + \frac{2\pi l}{\lambda} = \frac{3\pi l}{\lambda} + \pi. \tag{5.47}$$

Constructive interference will occur when the phase difference between these two wave components is a multiple of $2\pi$. Hence

$$\phi_B - \phi_F = \frac{2\pi l}{\lambda} + \pi = 2\pi n, \text{ where } n \in 1, 2, 3, \dots . \tag{5.48}$$

---

[8]Kossoff G. The effects of backing and matching on the performance of piezoelectric ceramic transducers. *IEEE Trans Son Ultrason* **1966** 13:20–30.

Solving for $l$ reveals that maximum signal output due to constructive interference of the two components occurs when

$$l = \frac{\lambda}{2}(2n - 1). \tag{5.49}$$

This is the classic half-wave resonator condition, seen throughout acoustics (*e.g.*, oscillations on a string, or resonances of a tube). Half-wave resonance occurs when there are symmetrical boundary conditions (*i.e.*, both surfaces of the resonator are bounded by a material that is acoustically hard, or both surfaces face an acoustically soft impedance). When a resonator has asymmetric boundary conditions (*e.g.*, one surface hard, one surface soft) the resonator exhibits quarter-wave resonance.[3,9] Due to the high characteristic acoustic impedance of most piezo-ceramics, it is difficult implementing a backing that has a much higher acoustic impedance than the piezo-electric element. However, piezo-polymers have much lower characteristic impedances than piezo-ceramics. This means that backing materials with acoustic impedances that are both greater than and less than that of the resonator are readily available. The effect of backing material on resonance mode is discussed by Brown[10] for the case of piezo-polymer transducers.

The other purpose of a transducer's backing is to provide damping. In the case of a highly resonant transducer, the amount of damping should be kept to an absolute minimum so that the piezo-electric element can resonate freely. This is normally accomplished by providing an open cavity behind the active element into which it can vibrate; this typed of undamped transducer is often referred to as an air-backed transducer. The resonant (undamped) data traces in Figures 5.3 and 5.4 were obtained from an air-backed device. A highly damped transducer will typically have a dense, high attenuation backing. Such backings are commonly produced by mixing fine metal, or alumina, powders into an epoxy resin and then casting the mixture directly onto the back of the piezo-electric materials. The damping provided by these sort of backings will reduce the amplitude of the produced ultrasonic signal but will increase the bandwidth ($BW$), as can be seen for the damped data trace in Figures 5.3 and 5.4. Damping often also serves to slightly reduce $f_{\rm c}$, which, combined with the increase in $BW$, results in an increase in $FBW$ as well.

### 5.4.3  Acoustic impedance matching

For most medical ultrasonic applications, the transducer produces waves that propagate into a medium with properties similar to water. The data in Table 5.4 shows that the acoustic impedance of the piezo-electric material is often much greater than water. Consider the interface between a hard PZT (35 MRayls) and water (1.5 MRayls). From (5.46) it can be seen that the piezo-ceramic/water

---

[9]Heuter TF, Bolt RH *Sonics*. New York: Wiley **1955**.

[10]Brown LF. The effects of material selection for backing and wear protection/quarter-wave matching of piezoelectric polymer ultrasound transducers. *Proc IEEE Ultrason Symp* **2000** 2:1029–1032.

interface has a reflection coefficient of $\frac{35-1.5}{35+1.5} = 91.8\%$. This means that more than 90% of the acoustic pressure produced within the piezo-ceramic is reflected at the interface with water. This reflected acoustic signal will "ring-around" inside the piezo-crystal, with less than 10% of the pressure able to propagate into the water each time it encounters the crystal/water interface.

This physical limitation is often overcome by using one or more thin layers applied to the front face of the transducer to provide a more gradual transition in acoustic impedance. The following discussion considers only one matching layer, but the reader is referred to Inoue *et al.*[11] and Desilets *et al.*[12] for a description of multiple matching layers. Combining (4.192)–(4.195) yields an expression for the reflection coefficient across a layer, $R_{\text{layer}}$, of

$$R_{\text{layer}} = \frac{B_1}{A_1} = \frac{(Z_1 + Z_2)(Z_2 - Z_3)e^{-j2k_2l} + (Z_1 - Z_2)(Z_2 + Z_3)}{(Z_1 + Z_2)(Z_2 + Z_3)e^{-j2k_2l} + (Z_1 - Z_2)(Z_2 - Z_3)}. \tag{5.50}$$

When the thickness of the layer, $l$, is a quarter of a wavelength long, *i.e.*,

$$l = \frac{\lambda_2}{4}, \tag{5.51}$$

it is found that

$$k_2 l = \frac{\pi}{2}. \tag{5.52}$$

This has the effect that the exponential term in (5.50) takes the value $-1 + j0$ and (5.50) reduces to

$$R_{\frac{\lambda}{4}} = \frac{Z_2^2 - Z_1 Z_3}{Z_2^2 + Z_1 Z_3}. \tag{5.53}$$

The numerator of (5.53) can be seen to vanish when

$$Z_2 = \sqrt{Z_1 Z_3}, \tag{5.54}$$

in which case $R_{\frac{\lambda}{4}} = 0$.

To recap, theoretically it is possible to achieve a reflection coefficient of zero (*i.e.*, complete transmission) if a matching layer is applied to a transducer's surface as long as:

- the matching layer is one quarter of a wavelength long, and

- the acoustic impedance of the matching layer is the geometric mean of the two media surrounding it.

Before proceeding, the limitations and assumptions used in the above analysis need to be clearly stated. Firstly, the boundary conditions used in the derivation

[11] Inoue T, Ohta M, Takahashi S. Design of ultrasonic transducers with multiple acoustic matching layers for medical application. *IEEE Trans Ultrason Ferroelectr Freq Control* **1987** 34:8–16.

[12] Desilets CS, Fraser JD, Kino GS. The design of efficient broad-band piezoelectric transducers. *IEEE Trans Son Ultrason* **1978** 25:115–125.

relate to fluid media. Both the matching layer and the piezo-electric element are solid media and therefore require a more comprehensive description involving shear wave propagation. Secondly, the underlying assumption is that the bounding media (materials 1 and 3) are of infinite extent. Whilst this may be true of the water (or water-like) material into which sound is propagating, it is certainly not true of the piezo-electric element. As was discussed in Section 5.4.1 the piezo-element is deliberately designed to be a resonant structure. Therefore the true impedance presented at the surface of the matching layer will be subject to loading by the acoustic backing applied on the piezo-element. Finally, no attempt has been made to account for dissipation and loss within the matching layer. For high frequency transducers such losses may be significant.

However, the above approach provides both a good illustration of the principle of matching layers and is also a close enough approximation for many practical applications, particularly at lower frequencies. If the above approach is found too simplistic, or if multiply matching layers are to be considered, then a more comprehensive theoretical model is required. Such discussion of the acoustics of layered media is beyond the scope of this text but the reader is referred to the excellent treatise by Brekhovskikh and Godin.[13]

The practicality of the two requirements of quarter-wave matching layers (listed above) also merits further discussion. The thickness criterion of the matching layer can be readily addressed either by careful machining of matching layers cast onto the surface of the piezo-electric element or by means of precision lapped layers that are subsequently applied to the transducer surface. However, whenever matching layers are separately applied, care must be taken to ensure that the bondline between the matching layer and the piezo-electric element is kept as thin as possible (ideally $< 1\,\mu$m).

The impedance of the matching layer material presents more of a challenge. Using (5.54) with the approximate values for the characteristic impedances of water and piezo-ceramic listed earlier suggests that a matching layer material should have an acoustic impedance of approximately 7.24 MRayls. Most polymers have acoustic impedances in the range 3–4.5 MRayls, whereas glasses are typically 14 MRayls and most metals exceed 20 MRayls. There are very few materials whose characteristic acoustic impedance is close to 7 MRayls. For this reason, it is not uncommon to simply use a rigid, castable polymer (*e.g.*, an epoxy resin) as a matching layer material under the assumption that its impedance will be "close enough". The validity of this assertion will now be examined.

Figure 5.7 plots (5.50) for two values of matching layer impedance. Even a small deviation from the target impedance of 7.24 MRayls results in minimum achievable reflection coefficient close to 0.2. If the best that can be achieved is 20% of the signal amplitude being reflected at the surface of the 6 MRayls matching layer, it is clear that using a simple epoxy (typically 4 MRayls) layer will have a much higher reflection coefficient. Clearly this will severely compromise transducer performance.

---

[13]Brekhovskikh L, Godin O. *Acoustics of layered media I*. Berlin: Springer **1990**.

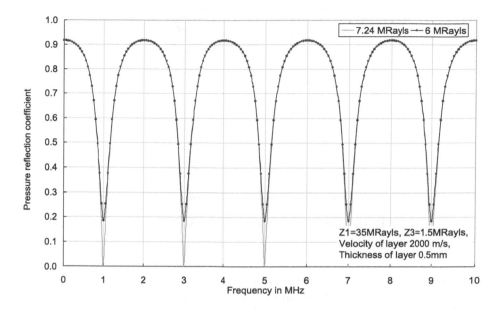

Figure 5.7: The variation of reflection coefficient from a matching layer as a function of its characteristic acoustic impedance.

To overcome the limitation on available materials, base resins such as epoxies or polyurethanes are often filled with high density (*e.g.*, tungsten) and/or high velocity (*e.g.*, alumina) powders to produce 0–3 composites with the required characteristics. The properties of these composites can be readily predicted with the Devaney[14] model. Several authors[15–17] have reported the use of such composites as matching layers.

## 5.4.4 Electrical impedance matching

As with all transmission lines, power flow into a load is maximised when the impedances of the load and the transmission line feeding it are matched. In the context of an ultrasonic system, the load is the transducer itself and the cable connecting it to the source of electrical signal is the transmission line. Many function generators and RF amplifiers have $50\,\Omega$ output. When devices like these are connected with a cable of characteristic impedance $50\,\Omega$ (such as

[14]Devaney AJ, Levine H. Effective elastic parameters of random composites. *Appl Phys Lett* **1980** 37:377–379.

[15]Wang H, Ritter TA, Cao W, Shung KK. High frequency properties of passive materials for ultrasonic transducers. *IEEE Trans Ultrason Ferroelectr Freq Control* **2001** 48:78–84.

[16]Rhee S, Ritter TA, Shung KK, Wang H, Cao W. Materials for acoustic matching in ultrasound transducers. In *Proc IEEE Ultrason Symp* **2001** 2:1051–1055.

[17]Zhou Q, Cha JH, Huang Y, Zhang R, Cao W, Shung KK. Alumina/epoxy nanocomposite matching layers for high-frequency ultrasound transducer application. *IEEE Trans Trans Ultrason Ferroelectr Freq Control* **2009** 56:213–219.

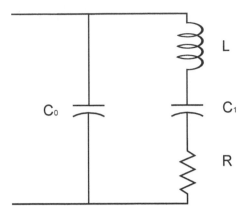

Figure 5.8: The Butterworth–Van Dyke equivalent circuit that is commonly used to model piezo-electric devices.

RG58 co-axial cable), there is optimum power transfer between devices and no undesirable standing waves within the cable. If this logic is then extended to consider the transducer, this too should be matched to the electrical impedance of the line used to drive it. Given the wide usage of 50-$\Omega$ systems, the remainder of this section will be based around matching to $50\,\Omega$. However, the ideas could equally be applied to other characteristic impedances.

> *CAUTIONARY NOTE*: A number of commercially available pulser–receiver systems have an electrical impedance that is not constant as a function of time. Typically these device have a low impedance whilst outputting the drive pulse, but a much higher impedance when receiving. Furthermore it is rare for either of the electrical impedances to be purely real (*i.e.*, it is often a complex valued impedance). Impedance matching a transducer to this type of pulser is therefore much harder (if not impossible).

When attempting to model the electrical impedance of a piezo-electric transducer, it is common to use a simple circuit representation of the device. A commonly used equivalent circuit is the Butterworth–Van Dyke model (see Figure 5.8) that places series inductive, capacitive and resistive components in parallel with another capacitor; this is the equivalent circuit recommended in ANSI/IEEE 176.[1] However, an alternative equivalent circuit has been suggested by Sherrit *et al.*[18] that more accurately represents the loss mechanisms associated with many of the physical constants. Recall the basic equations defining the impedances of simple electrical components:

$$Z_R = R; \qquad Z_L = j\omega L; \qquad Z_C = \frac{-j}{\omega C}. \qquad (5.55)$$

[18]Sherrit S, Wiederick HD, Mukherjee BK, Sayer M. An accurate equivalent circuit for the unloaded piezoelectric vibrator in the thickness mode. *J Phys D Appl Phys* **1997** 30:2354–2363.

The expressions for the impedance of an inductor $Z_L$ and of a capacitor $Z_C$ both contain $j$ ($j^2 = -1$). Therefore any simple equivalent circuit such as Figure 5.8 will have a complex-valued impedance. There are many common ways of displaying complex-valued impedance data:

- A pair of plots of resistance and reactance (real and imaginary parts of impedance).

- A pair of plots of impedance magnitude and phase.

- A pair of plots of conductance and susceptance (real and imaginary parts of admittance; admittance is the reciprocal of impedance).

- A Smith's Chart.

Note that only the last of these allows you to display complex data on a single plot, without the need to overlay two plots on the same axes. For this reason the Smith's Chart representation will be used within this section for the display of impedance data. For those readers unfamiliar with the Smith's Chart, a thorough explanation is provided by the chart's creator, Philip Smith;[19] however, a brief summary follows:

- A Smith's Chart has a horizontal axis running through the diameter of a circular plot. This horizontal line is the real axis and has $0\,\Omega$ on the left edge and $+\infty\,\Omega$ on the right edge. The resistance at the centre of the plot is the reference impedance $Z_0$ (assumed to be $50\,\Omega$ throughout this chapter).

- Any point below the real axis has a negative valued imaginary component and is capacitive. Similarly adding capacitance moves a data point towards the lower (capacitive) half of the chart.

- Any point above the real axis has a positive valued imaginary component and is inductive. Similarly adding inductance moves a data point towards the upper (inductive) half of the chart.

- The chart contains a series of concentric circles that are symmetric about the real axis and the all coincide on the right edge of the plot. These are circles of constant resistance.

- There are also two series of curves that run from the $+\infty$-$\Omega$ point on the real axis and pass into the upper and lower halves of the plot. These are lines of constant reactance.

- Each point on the Smith's Chart represents a unique complex-valued impedance. A plot of impedance as a function of frequency will result in curve on the chart.

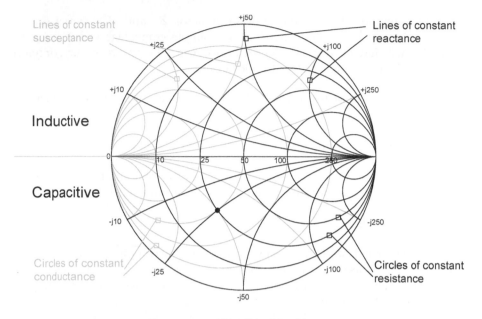

Figure 5.9: The Smith's Chart.

An example of a Smith's Chart can be found in Figure 5.9. Within this figure the lines of constant resistance and reactance are shown in black. A data point at $25 - j25\,\Omega$ has also been marked with a black dot; this will be referred to later. For convenience, lines of constant conductance and susceptance are also shown in pale grey; these lines are the mirror image of their impedance counterparts. The Smith's Chart is not only useful as a means of displaying data but it can assist in the design of impedance matching networks. Recall the equations for the combination of electrical impedances; a series combination of components has an impedance given by

$$Z_{\text{series}} = Z_1 + Z_2, \tag{5.56}$$

whereas a parallel combination is found through the equation

$$\frac{1}{Z_{\text{parallel}}} = \frac{1}{Z_1} + \frac{1}{Z_2}. \tag{5.57}$$

In practice (5.57) is more conveniently dealt with by transforming the problem from impedance $(Z)$ to admittance $(Y)$ since $Y = Z^{-1}$, therefore

$$Y_{\text{parallel}} = Y_1 + Y_2. \tag{5.58}$$

The implication of (5.56) and (5.58) in the context of the Smith's Chart is that it is easiest to add series components using the impedance version (black grid

---

[19]Smith PH. *Electronic Applications of the Smith Chart.* 2nd ed. Atlanta: Noble **2000**.

lines on Figure 5.9) whereas parallel components should be considered on an admittance version of the chart (grey grid lines on Figure 5.9).

For example, the addition of a purely capacitive series component will cause a counter-clockwise rotation of the marked data point along the 25-$\Omega$ line of constant resistance (since a pure capacitor makes no change to the resistance, but increases the negative reactive component). To bring the marked data point to the 50-$\Omega$ point on the real axis requires the addition of a parallel inductor. This will cause the data point to move towards the upper half of the chart along a line of constant conductance. The marked data point has an impedance $25 - j25\,\Omega$ so thus has an admittance of $0.02 + j0.02$ siemens. The value of the inductive correction needed is obtained from the admittance equivalent of $Z_L$ in (5.55). For convenience it is assumed that the operating frequency is 5 MHz.

$$Y_L = \frac{1}{j\omega L} \tag{5.59}$$

and therefore

$$L = \frac{1}{0.02 \times 2\pi \times 5 \times 10^6}\,\text{H} = 1.6\,\mu\text{H}. \tag{5.60}$$

In the above example, a single component could be used to match the electrical impedance of the transducer with the driving impedance. A matching scenario as simple as this is rare and often several reactive components are required. Under these conditions, the tolerances on the values of the individual components can become an issue. Furthermore very few components are purely reactive at typical ultrasound frequencies; most capacitors and inductors have stray resistances associated with them. Similarly, unless surface-mounted components are used, the electrode legs all act as stray inductors. These factors mean that a matching network made only from capacitors and inductors can be very sensitive to small changes in the component values. To achieve a more "robust" matching network, an alternative approach may be necessary.

Whilst it is well known that transformers can be used to step-up or step-down voltages, their ability to transform electrical impedances is sometimes overlooked. Consider a transformer with twice as many turns on its primary coil than it has on its secondary; this transformer has a turns ratio $n = 2$. A transformer of this nature would step-down electrical current by a factor of two. However, due to conservation-of-energy considerations, this same transducer would step-up voltage by a factor of two. Electrical impedance is the ratio of voltage and current, so that

$$Z_{\text{secondary}} = \frac{V_{\text{secondary}}}{I_{\text{secondary}}} = \frac{n\,V_{\text{primary}}}{\frac{1}{n}\,I_{\text{primary}}} = n^2\,Z_{\text{primary}}. \tag{5.61}$$

To exploit (5.61) in an impedance matching network, first consider the ratio of the resistive parts of both driving impedance and transducer and set this equal to $n^2$. For the example in Figure 5.9 this is $50/25 = 2 = n^2$, and thus the turns ratio $n = \sqrt{2}$. A primary winding with 7 turns and a secondary winding of 4 turns has a turns ratio of 1.4, which is a reasonable approximation to the required value. A simple and practical rule for ensuring the transformer is placed the correct way around in the matching circuit is

"place the largest number of turns against the largest impedance".

With this in mind, the secondary winding (with 4 turns) is connected to the piezo-electric element, whilst the primary winding (with 7 turns) faces the 50-$\Omega$ line that is driving the transducer. This arrangement will ensure the transducer now appears to have a resistive component of $50\,\Omega$ at the frequency of interest, although the reactive component will be non-zero. This is because the transformer will also have transformed any reactive component present, as well as introducing additional reactive contributions from the inductance of the coils and also inter-winding capacitances. The remaining reactive component can be corrected for with a single passive component (*i.e.*, capacitor or inductor). The value of this reactive correction is calculated in exactly the same manner as above.

When measuring the electrical impedance of a transducer (*e.g.*, with an impedance or vector network analyser) it should experience the same load conditions as would be seen in normal usage. Failure to do so will lead to an incorrect mechanical load on the transducer, which, due to the electro-mechanical coupling of a piezo-electric device, will appear as a change in the electrical properties of the device. Therefore if a transducer is designed to produce ultrasound that will propagate into water, its impedance should be measured with the face of the device in water. If the water vessel in which the transducer is measured is of limited size, care should also be taken to avoid acoustic reflections reaching the transducer. Such acoustic reflections will affect the electrical properties of the device, so the water vessel should ideally be lined with an acoustic absorber to eliminate such reflection artefacts.

Having discussed how the Smith's Chart may be used for impedance matching, it is now beneficial to consider what an optimum end point of the matching process should be.

## Narrowband matching

When matching a narrowband transducer, the aim is to have maximum power transfer into the transducer at the resonant frequency. At frequencies away from resonance it should become increasingly difficult to drive the transducer. A matching network compatible with this requirement would have its impedance spiral passing through $Z_0$ (= $50\,\Omega$) at the transducer's centre frequency $f_c$. The impedance at all other frequencies is likely to deviate from $Z_0$. The equation for the pressure reflection coefficient (5.46) has a direct electrical analogue. For a narrowband transducer matched to $Z_0$, the numerator of this expression is zero; thus the reflection coefficient is zero, and maximum power is transferred into the transducer.

## Broadband matching

A broadband transducer can be achieved if one is prepared to compromise optimum efficiency at resonance to ensure better efficiency outside resonance. On the Smith's Chart this could be seen by having impedance spiral being relatively close to $Z_0$ over a wide range of frequencies. This means that whilst the reflection coefficient will always be non-zero, its value will be small over an extended frequency range. In practice, arranging the resonant loop to be centred around $Z_0$ is an effective method to accomplish this goal.

## Practical example

To demonstrate the advantages of impedance matching, the following practical example is provided. The transducer shown here had a circular active element, diameter 9 mm. It was designed to operate with $f_c = 5$ MHz in water and have moderately damped response to provide a compromise between signal amplitude and bandwidth.

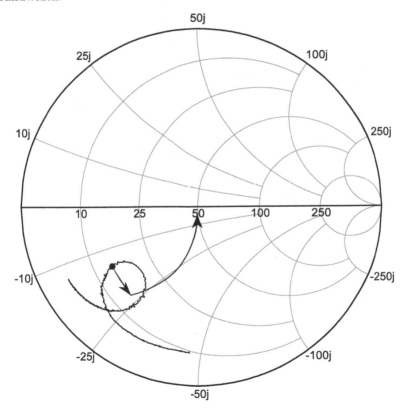

Figure 5.10: Electrical impedance before matching.

The initial electrical impedance of the transducer can be found in Figure 5.10. The centre frequency of the transducer has been marked on the figure with a black dot. Figure 5.10 has also been marked with two arrows to indicate the path of $f_c$ during impedance matching.

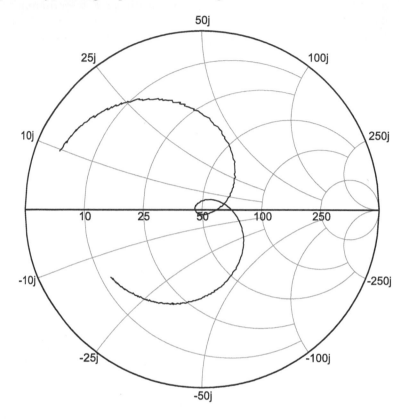

Figure 5.11: Electrical impedance after matching.

Initially the impedance at the operating frequency is brought onto the 0.02-siemens contour by the addition of a series capacitor. The remaining reactive component is then corrected by the addition of a parallel inductor. The impedance of the matched transducer can be seen in Figure 5.11, where an impedance loop around $50\,\Omega$ can be seen. The effect of impedance matching on the acoustic performance of this transducer is shown in Figures 5.12 and 5.13.

Figure 5.12 shows that the peak-to-peak amplitude of the acoustic signal has almost doubled due to impedance matching. Comparison of the spectra in Figure 5.13 reveals that $f_l$ and $f_u$ at a level 6 dB below the peak are wider apart for the matched transducer. This changes result in a increase in bandwidth from 2 MHz (unmatched) to 2.35 MHz (matched). In both cases $f_c$ is close to 5.3 MHz, so $FBW = 37\%$ for the unmatched transducer, increasing to $FBW = 43\%$ for the matched transducer.

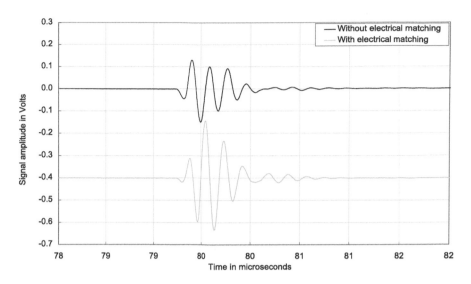

Figure 5.12: The effect of impedance matching on the temporal waveform from a 5-MHz, 9-mm diameter, PZT transducer.

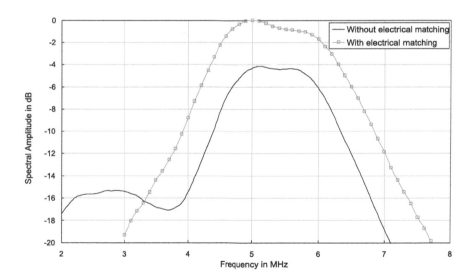

Figure 5.13: The effect of impedance matching on the spectrum of a 5-MHz, 9-mm diameter, PZT transducer.

# 6

# Radiated fields
Andrew Hurrell

The previous chapter was concerned solely with piezo-electric materials and the manner by which they are incorporated into an ultrasonic transducer. Having considered the internal structure of a transducer, the ultrasonic fields radiated by such a device are now presented. This chapter begins with an examination of continuous wave (CW) excitation, and specifically that from a circular plane piston source. Whilst the predicted field for this type of source is well known, it is of limited use in medical ultrasonics. Non-focussed physiotherapy devices are about the only devices that produce CW fields from a circular plane piston transducer. Despite these limitations, however, CW solutions are an instructive starting point as they introduce concepts that are useful in the study of more complex fields. The CW field produced by rectangular transducers is also presented.

Almost all devices capable of producing ultrasonic images do so with short pulses. To illustrate the differences with CW fields, pulsed excitation is considered separately in Section 6.2. High intensity therapeutic ultrasound extensively uses CW or quasi-CW excitation, but focussing is required to achieve the high ultrasonic intensities. Focussed fields will be covered separately in Section 6.3. This chapter finishes with an introduction to transducer arrays and how they can be used to provide both steered and focussed ultrasonic beams.

It should be noted that fields produced by many ultrasonic transducers are sufficiently complex that simple analytical descriptions do not exist. In these cases, the only option is to employ numerical techniques (such as a finite element or finite difference methods) to model the problem. However, numerical solutions suffer from the disadvantage that they only solve for one specific set of initial and boundary conditions, and may not necessarily yield a result that is generally applicable. In order to maintain generality, the field prediction methods presented in this chapter are based on analytic solutions.

## 6.1 Continuous wave excitation

The most widely studied model of ultrasonic radiation is that of a plane piston, radiating continuously at a single frequency, surrounded by a rigid baffle. Figure 6.1 contains a diagrammatic representation of an arbitrarily shaped piston in an infinite baffle. Within this illustration, the measurement point is at coordinates $(x, y, z)$, an arbitrary point on the source is at coordinates $(x', y', 0)$, and the vector between them is $\vec{r}$. The acoustic field at the measurement point expressed in terms of the velocity potential $\phi(x, y, z)$ is provided by the Rayleigh

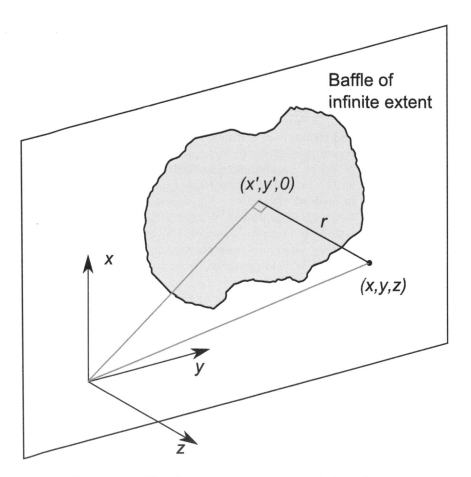

Figure 6.1: The plane piston source in an infinite baffle.

integral:

$$\phi(x, y, z) = \int_S \frac{v_0 e^{j(\omega t - kr)}}{2\pi r} \, dS, \tag{6.1}$$

where $dS = dx' \, dy'$ and $v_0$ is the amplitude of the particle velocity. For convenience, the temporal dependence $e^{j\omega t}$ will not be shown explicitly although its presence is implied throughout. This integral is simply an expression of Huygens' principle and states that the acoustic field at measurement point $(x, y, z)$ is the sum of all spherically radiating point sources that go to make up the surface $S$. The Rayleigh integral is simple in form, but much harder to solve analytically, without some simplifying assumptions.

## 6.1.1 Circular plane piston in a rigid baffle

### The full field

A common simplification of the geometry shown in Figure 6.1 is to require that the source piston is circular[1−4] For this case it is convenient to address the problem in a cylindrical coordinate system $(\rho, \eta, z)$ where $\rho^2 = x^2 + y^2$ and $\eta = \arctan\left(\frac{x}{y}\right)$. A rigorous solution of the Rayleigh integral is beyond the scope of this chapter and the reader is referred to the work of Hutchins $et~al.$[2] for a comprehensive mathematical derivation. However, the result obtained is worthy of comment and is given by

$$\phi(\rho, z) = \frac{-j \, v_0}{k} \left( \begin{bmatrix} 1 & : \rho < a \\ 1/2 & : \rho = a \\ 0 & : \rho > a \end{bmatrix} e^{-jkz} + \frac{1}{\pi} \int_0^\pi e^{-jkr} \frac{a\rho \cos\psi - a^2}{a^2 + \rho^2 - 2a\rho \cos\psi} d\psi \right), \tag{6.2}$$

where $\psi$ is a variable of integration, $a$ is the radius of the circular piston, and $r^2 = z^2 + a^2 + \rho^2 - 2a\rho \cos\psi$. The first term in (6.2) is a plane wave that only makes a contribution to the field when $\rho \leq a$. The second term is an edge wave that radiates from all points on the circumference of the source. As this edge wave propagates it interacts with both the plane wave component, and with the edge wave components from elsewhere on the circumference of the disc. These interactions produce an interference pattern whose amplitude can vary greatly as a function of spatial position. An example of the complex spatial distribution of the acoustic field radiated from a CW circular plane piston can be found in Figure 6.2.

---

[1] Archer-Hall JA, Gee D. A single integral computer method for axisymmetric transducers with various boundary conditions. $NDT~Int$ **1980** 13:95–101.

[2] Hutchins DA, Mair HD, Puhach PA, Osei AJ. Continuous-wave pressure fields of ultrasonic transducers. $J~Acoust~Soc~Am$ **1986** 80:1–12.

[3] Goldstein A. Steady state unfocused circular aperture beam patterns in nonattenuating and attenuating fluids. $J~Acoust~Soc~Am$ **2004** 115:99–110.

[4] Mast TD, Yu F. Simplified expansions for radiation from a baffled circular piston. $J~Acoust~Soc~Am$ **2005** 118:3457–3464.

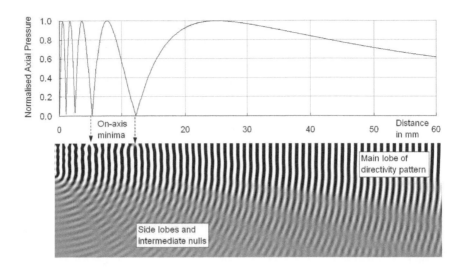

Figure 6.2: The axial acoustic pressure profile (*upper*) and a greyscale image of a radial slice of the whole acoustic pressure field (*lower*) produced by a circular plane piston transducer (diameter 10 mm) driven with a continuous 1.5-MHz sinusoid. Black indicates positive acoustic pressure, white indicates negative acoustic pressure. Greyscale image produced with AFiDS Suite (AMH Consulting, Poole, UK).

**The axial field**

The general solution provided by (6.2) can be simplified to yield the field along the acoustic axis of the source. On the acoustic axis, the radial coordinate $\rho = 0$ and therefore the velocity potential becomes

$$\phi(0, z) = \frac{-j\, v_0}{k} \left( e^{-jkz} - e^{-jk\sqrt{z^2+a^2}} \right). \tag{6.3}$$

The relationship between acoustic pressure and velocity potential is given by

$$p = -\rho_0 \frac{\partial \phi}{\partial t}, \tag{6.4}$$

where $\rho_0$ is the density of the medium within which the wave is propagating. Given the form of the implicit temporal dependence, $e^{j\omega t}$, a differentiation with respect to time can be achieved by multiplying by $j\omega$. Inserting (6.3) into (6.4) results in the following expression for the axial acoustic pressure:

$$p(0, z) = \frac{\omega \rho_0 v_0}{k} e^{j\omega t} \left( e^{-jkz} - e^{-jk\sqrt{z^2+a^2}} \right). \tag{6.5}$$

The magnitude of the term in parentheses in (6.5) defines the magnitude of the spatial distribution of the "on-axis" wave. Expanding the exponentials inside the parentheses with Euler's relation and combining real and imaginary parts yields a purely trigonometric expression, *viz.*

$$\cos kz - \cos \sqrt{z^2 + a^2} - j \left( \sin kz - \sin \sqrt{z^2 + a^2} \right). \tag{6.6}$$

Multiplying this expression by its complex conjugate and then applying several trigonometric simplifications produces

$$|p(0, z)| = 2Z_0 v_0 \left| \sin \left[ \frac{k}{2} (\sqrt{z^2 + a^2} - z) \right] \right|, \tag{6.7}$$

where $Z_0$ is the acoustic impedance of the material.

An example of the axial profile of a CW piston obtained from (6.7) can be found in Figure 6.2. This plot shows the presence of a series of minima and maxima along the acoustic axis. These peaks and troughs are caused by the interference between the principal plane wave from the transducer's surface and the wave radiating from the edges. Due to the high symmetry of a circular source, the edge waves from every point on the circumference will all arrive at a point on axis with the same phase. This can result in complete destructive interference with the plane wave.

To find the axial positions of these extrema first observe that for $z > a$ the square root in (6.7) can be replaced with its binomial expansion. Retaining only the first two terms of the expansion gives

$$|p(0, z)| = 2Z_0 v_0 \left| \sin \left( \frac{ka^2}{4z} \right) \right|. \tag{6.8}$$

From (6.8) it can be seen that destructive interference corresponds to locations where the sine function takes the value of 0. Similarly, the sine function takes the value 1 when the interference is constructive. These $z$-locations are found by setting

$$\frac{ka^2}{4z} = \frac{n\pi}{2} \left\{ \begin{array}{ll} n \text{ is odd:} & \text{maxima} \\ n \text{ is even:} & \text{minima} \end{array} \right\}. \tag{6.9}$$

The furthest maximum from the transducer face is found by setting $n = 1$ in (6.9). Solving for $z$ reveals that this last axial maximum is found at a distance

$$z = \frac{a^2}{\lambda}, \tag{6.10}$$

where $\lambda$ is the wavelength. This distance is often referred to as the near-field/far-field transition point, or simply the "transition distance". As pointed out by Goldstein,[3] the term "far-field" should only really refer to distances sufficiently far from the transducer where the amplitude has a $\frac{1}{r}$ dependence. Whilst the axial pressure amplitude is monotonically decreasing at distances beyond (6.10) the true "far field" is often much further away.

Before proceeding, it is essential to stress the limitations on the use of (6.10). All too often (6.10) is used to predict the position of the last axial maximum

128

without regard to how the transducer is being used. The entire derivation above
was based around a CW source of excitation. When a transducer is driven with a
transient signal there may be little (if any) opportunity for interaction between
plane and edge wave components since they are temporally separated. This
can result in the last axial maximum being much closer to the transducer than
predicted by (6.10). Transient excitation will be further discussed in Section
6.2.

**The directional response**

For points that are truly in the far field of a circular plane piston it is also
possible to approximate the directional response without needing to evaluate the
full field description of (6.2). A comprehensive derivation of this approximation
is provided by Pierce[5] but the key points are summarised below.

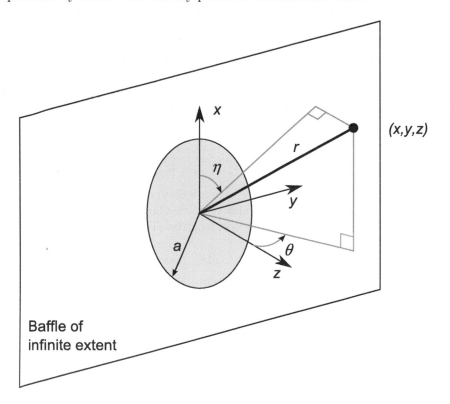

Figure 6.3: A circular plane piston source in an infinite baffle.

For a circular plane piston the geometry of Figure 6.1 is modified such that
the coordinate system is as shown in Figure 6.3. It can be shown that when $r$

[5]Pierce AD. *Acoustics: An Introduction to Its Physical Principles and Applications*. 1994
ed. Melville: Acoustical Society of America **1994**.

is approximated with the first two terms of its binomial expansion, and when $r \gg a$, the acoustic pressure $p(r, \theta, t)$ is given by

$$p(r, \theta) = -j \frac{P_0 e^{-jkr}}{2\pi r} \int_{\rho=0}^{\rho=a} \left( \int_{\eta=0}^{\eta=2\pi} e^{jk\rho \sin \theta \cos \eta} \, d\eta \right) \rho \, d\rho, \tag{6.11}$$

where $\theta$ is the out-of-plane angle, $\eta$ is the in-plane angle and $P_0$ is the pressure amplitude. The zeroth order Bessel function of the first kind is defined as

$$J_0(\sigma) = \frac{1}{2\pi} \int_0^{2\pi} e^{j\sigma \cos \zeta} d\zeta. \tag{6.12}$$

Substituting (6.12) into (6.11) yields

$$p(r, \theta) = -j \frac{P_0 e^{-jkr}}{2\pi r} \int_{\rho=0}^{\rho=a} 2\pi J_0(k\rho \sin \theta) \rho \, d\rho. \tag{6.13}$$

However (6.13) can be further simplified with the relationship

$$\int \sigma J_0(\sigma) \, d\sigma = \sigma J_1(\sigma). \tag{6.14}$$

To exploit (6.14) a change of variable $\sigma = k\rho \sin(\theta)$ is needed and thus

$$\int_{\rho=0}^{\rho=a} J_0(k\rho \sin \theta) \rho d\rho = \frac{1}{k^2 \sin^2 \theta} \int_{\sigma=0}^{\sigma=ka \sin \theta} \sigma J_0(\sigma) \, d\sigma. \tag{6.15}$$

Combining (6.14) with (6.15), substituting into (6.13), and simplifying gives a final form of

$$p(r, \theta) = \frac{-ja^2 P_0}{2} \frac{e^{-jkr}}{r} D(\theta, k), \tag{6.16}$$

where $D(\theta, k)$ (often referred to as the far-field directivity pattern of a circular plane piston transducer) is given by

$$D(\theta, k) = \left[ \frac{2J_1(ka \sin \theta)}{ka \sin \theta} \right]. \tag{6.17}$$

Figure 6.4 shows a plot of (6.17) for a transducer of radius 5 mm, operating at both 0.5 and 1.5 MHz in water. The higher frequency case corresponds to configuration displayed in Figure 6.2.

In his 1966 paper Williams[6] observed that (6.17) is only valid at large distances from the source. This assertion was subsequently quantified by Goldstein,[3] who compared the shape of far-field directional responses provided by (6.16) with lateral profiles derived from (6.2) and found that the two were identical at 6.41 times the transition distance. He also considered the position of the first minimum of (6.16) and the first minima of a lateral profile determined

[6] Williams Jr AO. Medium- and far-field expressions for velocity potential of circular plane piston. *J Acoust Soc Am* **1966** 39:1142–1144.

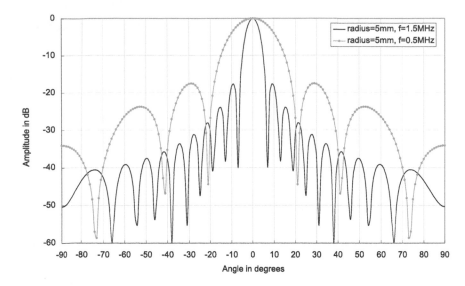

Figure 6.4: The approximation to the far-field directivity function of a circular plane piston as given by (6.17).

from a full beam expression such as (6.2). He showed that the positions of these minima were the same at 1.15 times the transition distance.

These two cases are important because models based upon (6.17) are often used to estimate the effective radius of an experimentally measured, plane piston source. A determination of effective radius based upon the position of the first lateral minimum is likely to be relatively accurate at distances of only 1.15 times the transition distance. However, it is not uncommon to evaluate effective radius by comparing the experimentally determined −3-dB and −6-dB widths of the main lobe with similar values from (6.17). This latter case is dependant upon the shape of the main lobe and should therefore only be undertaken with measurements taken at much greater axial distances.

Therefore as a general rule when determining effective radius from the width of the main lobe, unless axial distances are sufficiently great (*e.g.*, 6×transition distance), reliance on the approximation provided by (6.17) may be unwise. Instead a full description of the acoustic field radiated from the circular plane piston should be used. Equation (6.2) can be readily evaluated with numerical integration techniques or other solutions of comparable accuracy such as that of Mast and Yu[4] or Mellow.[7]

---

[7]Mellow T. On the sound field of a resilient disk in free space. *J Acoust Soc Am* **2008** 123:1880–1891.

## 6.1.2 Rectangular plane piston in an infinite baffle

### The full field

Whilst circular plane pistons provide a useful, and comparatively simple, model against which single element transducers can be compared, they are of little benefit for non-circular geometries. The vast majority of diagnostic imaging arrays use rectangular elements in either LE or WE modes. Prediction of the field produced by these transducers requires the solution of the Rayleigh integral for CW radiation from a rectangular plane piston.[8,9]

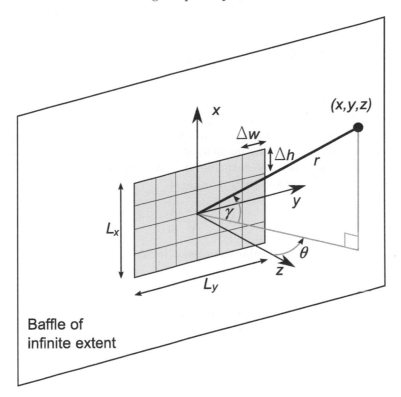

Figure 6.5: A rectangular plane piston source in an infinite baffle.

As with the case of the circular plane piston transducer, the reader is referred to the literature for the detailed mathematical derivations, but the method of Ocheltree and Frizzel[8] is briefly described here. For rectangular sources a Cartesian geometry is most appropriate. The rectangular source of dimensions $L_x$ and $L_y$ (as shown in Figure 6.5) is subdivided into $N$ smaller rectangular regions. Each region is too large to be considered as a Huygens point source,

[8]Ocheltree KB, Frizzel LA. Sound field calculation for rectangular sources. *IEEE Trans Ultrason Ferroelectr Freq Control* **1989** 36:242–248.

[9]Ding D, Zhang Y, Liu J. Some extensions of the Gaussian beam expansion: radiation fields of the rectangular and the elliptical transducer. *J Acoust Soc Am* **2003** 113:3043–3048.

but is still small enough to permit some simplifying assumptions. The centre of the $n^{\text{th}}$ region is $(x_n, y_n)$ and all regions have a width and height of $\Delta w$ and $\Delta h$ respectively. The elemental area is thus $\Delta A = \Delta w \Delta h$. For an arbitrary point in the field $(x, y, z)$ the following distances can be defined

$$\hat{x}_n = x - x_n \,;$$

$$\hat{y}_n = y - y_n \,; \tag{6.18}$$

$$r = \sqrt{\hat{x}_n^2 + \hat{y}_n^2 + z^2} \,.$$

The CW pressure from a rectangular source can then be defined as

$$p(x, y, z) = \frac{j\rho \Delta A c}{\lambda} \sum_{n=1}^{N} \left[ \frac{u_n e^{-jkr}}{r} \operatorname{sinc}\left( \frac{k\hat{x}_n \Delta w}{2r} \right) \operatorname{sinc}\left( \frac{k\hat{y}_n \Delta h}{2r} \right) \right], \tag{6.19}$$

where $u_n$ is the surface velocity of element $n$, and $\operatorname{sinc}(x) = \frac{\sin(x)}{x}$. The assumptions used in the derivation above require that the measurement point $(x, y, z)$ be in the far field of each rectangular subregions (NB: this does not require that the measurement point is in the far field of the overall source). This assumption can effectively be satisfied by ensuring that $\Delta w$ and $\Delta h$ are kept sufficiently small. Ocheltree and Frizzel[8] used a value of

$$\Delta w = \sqrt{\frac{4\lambda z}{10}} \tag{6.20}$$

for their examples, with a similar expression for $\Delta h$.

### The axial field

Equation (6.19) has been used to calculate the axial acoustic pressure from a square piston transducer whose side length matches the diameter of the circular plane piston of Figure 6.2. The results for the rectangular plane piston are displayed in Figure 6.6. Note that for the square transducer, the last axial maximum is much further away than for the circular case. Furthermore, although

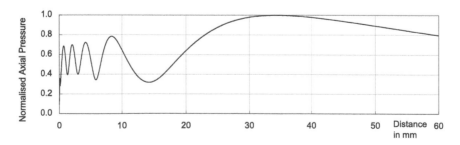

Figure 6.6: The axial acoustic pressure profile produced by a square plane piston transducer (side length 10 mm) driven with a continuous 1.5-MHz sinusoid.

the on-axis pressure exhibits maxima and minima, the range of their amplitude variation is much less than the circular case. As before, the explanation of these features requires an understanding of the edge wave behaviour. The symmetry of a circular source ensures that the path length to an on-axis point is the same from all points on the circumference. This enables complete destructive interference of plane and edge wave components. In contrast, square (or rectangular) sources have a very much restricted symmetry. Consider an on-axis measurement point. The path length from this point to the corner of the source is larger than the distance from the measurement point to the middle of one side. This range of path lengths prevents all edge wave component arriving on-axis at the same time and thus complete destructive interference with the plane wave does not occur.

### The directional response

The directional response of a rectangular radiator has already been implied in the expression of the full pressure field. Recall that (6.19) is a sum of sinc functions each of which relates to the far field of one of the rectangular sub-regions. A familiar concept in optics is that the far-field radiation pattern of an aperture is simply its spatial Fourier transform.[10] A boxcar (or rectangle) function is defined as one whose value is zero every where except for a single, finite interval within which the value is constant. Therefore the velocity profile of a plane rectangular source takes the shape of a boxcar function. It is well known that the Fourier transform of a boxcar function is a sinc function. For a rectangular plane piston the directional response is simply the product of two sinc functions[11] (one in each of the $x$ and $y$ directions) and is given by

$$p(r, \theta, \gamma) = \frac{-j4L_x L_y P_0}{2\pi} \frac{e^{-jkr}}{r} \mathrm{sinc}\left(\frac{kL_y}{2} \cos\gamma \sin\theta\right) \mathrm{sinc}\left(\frac{kL_x}{2} \sin\gamma\right). \quad (6.21)$$

Observe that (6.21) can be considered as the limiting case of (6.19) when the source is only divided into one subregion.

## 6.2 Transient excitation

In his review of transient radiation, Harris[12] discusses the wide range of analytical techniques available to the field patterns not due to CW excitation. One of the most widely used methods for predicting transient fields was developed by Stepanishen.[13] The Stepanishen method relies upon a Green's function ap-

---

[10]Goodman JW. *Introduction to Fourier optics*. 3rd ed. Englewood: Roberts **2005**.

[11]Molloy CT. Calculation of the directivity index for various types of radiators. *J Acoust Soc Am* **1948** 20:387–405.

[12]Harris GR. Review of transient field theory for a baffled planar piston. *J Acoust Soc Am* **1981** 70:10–20.

[13]Stepanishen PR. The time-dependent force and radiation impedance on a piston in a rigid infinite planar baffle. *J Acoust Soc Am* **1971** 49:841–849.

proach to solve the Rayleigh integral (albeit expressed in terms of the acoustic velocity potential).

For a piston transducer (*i.e.*, one whose velocity distribution $v(t)$ is uniform across its surface $S$), the velocity potential can be expressed as

$$\phi(\vec{r}, t) = \int_0^t v(t_0) \, dt_0 \int_S g(\vec{r}, t | \vec{r_0}, t_0) \, dS, \qquad (6.22)$$

where $g(\vec{r}, t | \vec{r_0}, t_0)$ is the appropriate Green's function. To emphasize the difference between CW and transient cases, the temporal dependence of particle velocity $v(t)$ will be explicitly retained in all equations. For this type of problem, the Green's function has been previously solved[14] and is given by

$$g(\vec{r}, t | \vec{r_0}, t_0) = \frac{\delta\left(t - t_0 - \frac{|\vec{r} - \vec{r_0}|}{c}\right)}{2\pi |\vec{r} - \vec{r_0}|}. \qquad (6.23)$$

Substitution of (6.23) into the second integral of (6.22) allows the definition of a spatial impulse response function $h(\vec{r}, t - t_0)$ as

$$h(\vec{r}, t - t_0) = \int_S \frac{\delta\left(t - t_0 - \frac{|\vec{r} - \vec{r_0}|}{c}\right)}{2\pi |\vec{r} - \vec{r_0}|} \, dS. \qquad (6.24)$$

The solution for the velocity potential is therefore

$$\begin{aligned}
\phi(\vec{r}, t) &= \int_0^t v(t_0) h(\vec{r}, t - t_0) \, dt_0 \\
&= v(t) * h(\vec{r}, t),
\end{aligned} \qquad (6.25)$$

where $*$ denotes convolution. Thus the problem of finding the field produced by transient excitation of an arbitrary piston shape has been reduced to finding the relevant spatial impulse response function, and convolving this with the source waveform.

## 6.2.1 Transient radiation from a circular plane piston

As for the CW case, the geometry of the circular plane piston problem is best expressed in a cylindrical coordinate system $(\rho, \eta, z)$, with a circular piston radius $a$. The solution for the spatial impulse response function from a circular piston has been shown by various authors[13,15] to be expressed in two parts. For field points within the projection of the circular piston (*i.e.*, $\rho < a$)

$$\begin{aligned}
h(\vec{r}, t) &= 0 & &: t < t_1, t > t_3 \\
&= c & &: t_1 < t < t_2 \qquad (6.26) \\
&= \frac{c}{2\pi} \Omega(\vec{r}, t) & &: t_2 < t < t_3
\end{aligned}$$

[14]Morse PM, Ingard KU. *Theoretical acoustics*. Princeton: Princeton University Press **1986**.

[15]Lockwood JC, Willette JG. High-speed method for computing the exact solution for the pressure variations in the nearfield of a baffled piston. *J Acoust Soc Am* **1973** 53:735–741.

and for points outside this projection (*i.e.,.* $\rho > a$)

$$h(\vec{r}, t) \quad = \quad 0 \qquad\qquad : t < t_2, t > t_3$$
$$= \quad \frac{c}{2\pi} \Omega(\vec{r}, t) \quad : t_2 < t < t_3 \tag{6.27}$$

where $t_1 = \frac{z}{c}$ and is the shortest propagation time from a point on the surface of the transducer to the measurement point. In comparison with the CW case, this is the arrival time of plane wave component from the transducer's surface. Similarly $t_2 = \sqrt{z^2 + (\rho - a)^2}/c$ and $t_3 = \sqrt{z^2 + (\rho + a)^2}/c$ are the arrival times of the edge wave components from the points on the circumference that are nearest and farthest to the measurement point. Finally the function $\Omega(\vec{r}, t)$ is given by

$$\Omega(\vec{r}, t) = 2 \arccos \left( \frac{(ct)^2 - z^2 + x^2 - a^2}{2x\sqrt{(ct)^2 - z^2}} \right). \tag{6.28}$$

The spatial impulse response function of (6.26) and (6.27) simply describes the interaction of plane and edge wave components at the measurement point. Inserting the spatial impulse response function into (6.25) yields the velocity potential in terms of plane and edge wave interactions in much the same way that (6.2) did for CW excitation of a circular piston source.

Figure 6.7 shows the radial field produced by the same circular piston source as Figure 6.2, but this time driven by a single cycle of a sinusoid at 1.5 MHz, to which a Gaussian window has been applied.

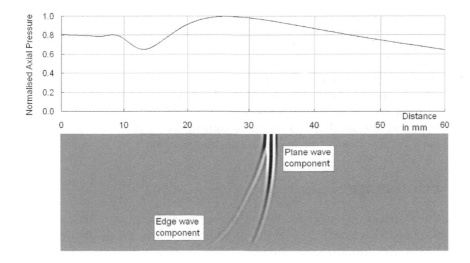

Figure 6.7: The axial acoustic pressure profile (upper) and a greyscale map of a radial slice of the acoustic pressure field (lower) produced by a circular plane piston transducer (diameter 10 mm) driven with a single cycle Gaussian pulse with centre frequency 1.5 MHz. Greyscale image produced with AFiDS Suite (AMH Consulting, Poole, UK).

Comparison of the axial profiles of Figures 6.2 and 6.7 shows that the transient case has far fewer on-axis minima than the CW case; furthermore these minima are broader and shallower than for CW excitation. It is also worth noting that the position of the last axial maximum is different from that expected when the transducer is driven continuously. Notice also from the greyscale map the complete absence of off-axis side lobes and nulls from Figure 6.7. Due to the finite duration of the pulsed signal there is no opportunity for the interference between edge and plane wave components that would lead to such complex off-axis behaviour. This figure clearly illustrates the differences between a CW field compared with one from a transducer subject to transient excitation.

It is not uncommon to see expressions in application notes that are derived from either (6.17) or (6.10) to characterise the field from a circular piston source (*e.g.*, to find the beam spread angle or the effective diameter of the source). However, in many cases it is also clear that the intended application involves using the transducer with a transient signal. Clearly the CW assumption underlying the theoretical expression is in contrast with the intended practical use of the transducer. The user should clearly recognise and understand the differences that can arise depending on how a transducer is driven.

### 6.2.2   Transient radiation from a rectangular plane piston

The reduced symmetry of a rectangular piston, relative to a circular one, leads to a somewhat more complicated expression for $h(\vec{r}, t)$. As was seen for circular pistons, the propagation times between various points on the edge of the rectangular piston is a critical component in the form of the final expression of the spatial impulse response. In the case of the circular piston three critical times $t_1$–$t_3$ were identified and the spatial impulse response function had only three non-zero components within (6.26) and (6.27). The Lockwood and Willette[15] description of the spatial impulse response function of a rectangular plane piston identifies 8 separate time delays and this leads to a solution containing 18 separate components. These have been conveniently tabulated by San Emeterio and Ullate[16] and the reader is advised to consult this paper (and references therein) for a comprehensive examination of this problem.

## 6.3   Focussing

Many medical ultrasonic systems employ one or more forms of focussing, and several methods of focussing will be presented later. All the methods discussed in this section are applicable to single element transducers; multi-element focussing methods (such as electronically phased arrays) are reserved for Section 6.4.

The simple representation of a focussed field within Figure 6.8 shows both the curvature of the wavefronts and arrows indicating the direction of travel

---

[16]San Emeterio JL, Ullate LG. Diffraction impulse response of rectangular transducers. *J Acoust Soc Am* **1992** 65:651–662.

at various locations. The wavefronts propagate forward until they converge at the focal region, before diverging post-focus. The lateral extents of the acoustic beam are clearly much smaller within the focal region than they are at source. This aspect of focussing can be to minimise the lateral spread and is often used in ultrasonic imaging systems to produce collimated beams of ultrasound. Note

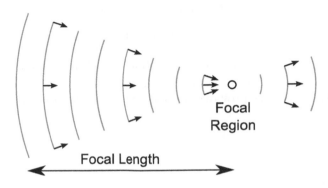

Figure 6.8: The basics of focussing.

also from Figure 6.8 the arrows showing direction of wavefront travel become much closer as the focal region is approached. This is a graphical illustration of the concentration of acoustic energy at the focus, and the corresponding increase of acoustic pressure and intensity within the focal region. The pressure focal gain $G_p$ of a focussed ultrasonic transducer system is defined as

$$G_p = \frac{\text{Pressure at transducer focus}}{\text{Average pressure at transducer surface}}. \tag{6.29}$$

HIFU devices often incorporate shaped piezo-ceramic elements (sometimes in combination with phased array technologies). These devices are designed to exploit focussing to concentrate acoustic energy into a small area, thereby increasing the ultrasound induced heating and subsequent tissue necrosis.

In practice real focussed fields are more complex than is indicated by Figure 6.8. As was seen previously, any radiating transducer surface will have plane wave and edge wave components. Circularly radiating edge waves will also be present for focussed ultrasound systems and thus, as before, there can be complex field structures arising from the interference of these different components. For this reason, the position of the focal depth may not necessarily coincide with simple estimators of focal depth (for example, the radius of curvature of a focussed radiator).

Focussed systems are often described in terms of the $F$-number (sometimes referred to as $Fn$ or $F\#$) and is defined as

$$F\text{-number} = \frac{z_f}{AW}, \tag{6.30}$$

where $z_f$ is the focal depth of the transducer and $AW$ is the aperture width. For a circularly symmetric transducer system $AW$ is simply the transducer diameter. However, for a rectangular transducer the length and width of the active element will not generally be equal and there may be different focal depths associated with each of these dimensions. Therefore it may be necessary to define a separate $F$-number for the length and width foci.

## 6.3.1 Shaped piezo-electric elements

The simplest way of producing a focussed ultrasonic field is to use a piezo-electric material that has an inherent radius of curvature. The case of a circular focussed radiator was first considered by O'Neil,[17] but has subsequently been addressed by several authors, including Kossoff,[18] Lucas and Muir,[19] and Goldstein.[20] The last of these papers provides a good review of the literature of focussed sources.

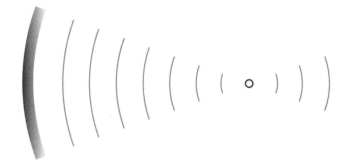

Figure 6.9: Use of a shaped piezo-electric element to produce curved wavefronts.

The magnitude of the axial acoustic pressure of a spherically focussed circular piston transducer (comparable to (6.8) for plane sources) is given by

$$|p(0, z)| = Z_0 v_0 \frac{2A}{A - z} \sin\left\{\frac{k}{2}[B(z) - z]\right\}, \tag{6.31}$$

where $B(z) = \sqrt{z^2 + 2b(A - z)}$, $b = A - \sqrt{A^2 - a^2}$, and $A$ is the radius of curvature of the spherical bowl. For a spherical bowl subject to CW excitation, the pressure focal gain can be approximated as

$$G_\mathrm{p} = \frac{\pi a^2}{F_\mathrm{geo}\lambda}, \tag{6.32}$$

where $F_\mathrm{geo}$ is the geometric focal length.

[17]O'Neil HT. Theory of focusing radiators. *J Acoust Soc Am* **1949** 21:516–526.

[18]Kossoff G. Analysis of focusing action of spherically curved transducers. *Ultrasound Med Biol* **1979** 5:359–365.

[19]Lucas BG, Muir TG. The field of a focusing source. *J Acoust Soc Am* **1982** 72:1289–1296.

[20]Goldstein A. Steady state spherically focused, circular aperture beam patterns. *Ultrasound Med Biol* **2006** 32:1441-1458.

## 6.3.2 Acoustic lenses

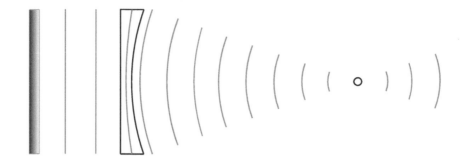

Figure 6.10: Use of an acoustic lens to induce curvature to previously parallel wavefronts.

When considering acoustic lenses it is instructive to make a comparison with optics. The speed of light in air is very close to the fundamental upper limiting value of the speed of light in a vacuum. Therefore in practice, any optical components must be made from materials in which the speed of light is lower than that in the surrounding medium. Consequently a glass plate with one planar and one convex surface (a plano-convex lens) will always form a converging lens.

In contrast to the optics case, materials used to construct an acoustic lens could have acoustic speeds that are either faster or slower than those in the surrounding materials. Consider propagation of an ultrasound wave in water (speed of sound $1480\,\mathrm{m\,s^{-1}}$). Many epoxy resins have sound speeds in excess of $2000\,\mathrm{m\,s^{-1}}$, whilst the majority of metals have sound speeds greater than $3000\,\mathrm{m\,s^{-1}}$. In contrast many silicone rubbers have a speed of sound below $1000\,\mathrm{m\,s^{-1}}$. Clearly then, an acoustic plano-convex lens could be either converging or diverging, depending on the velocity of sound in the material from which it is made.

Another important difference between optics and acoustics arises from the vast difference in the speeds of light and sound. The transit time of light across an optical lens is negligible, but the same cannot be said for ultrasonics systems. Consider the acoustic lens shown in Figure 6.11, where the material of the lens is faster than the surrounding medium. In the time taken by a wave at the edge of the lens to travel a distance $Z_1$, a wave in water at the centre will have travelled a distance $Z_2$. Therefore the effective radius of curvature of the wavefront leaving the lens is shown by the dashed grey line, which is much shallower than the geometric radius of curvature. Consequently the effective focal length (also shown with a dashed grey line) is $\Delta f$ larger than the geometric focal length.

A simple geometric calculation could be conducted to estimate the increase in focal length as a function of lens properties and dimensions. However, this would not account for phenomena such as edge/plane wave interactions or secondary diffraction of the edge wave. In practice, when designing a transducer that

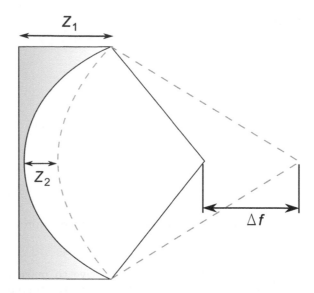

Figure 6.11: Effective focal length of a lens: the solid black line indicates the geometric focus, the dashed grey line the effective focus.

incorporates an acoustic lens, a numerical model is often used to predict the true position of the acoustic focus.

## 6.4 Transducer arrays

The logical extension from a single element transducer is to have an array of transducer elements. By dividing a piezo-electric material into smaller subregions it is possible to reduce its effective capacitance. This in turn reduces the electrical impedance of the piezo-electric element, and it is thus easier to electrically drive the transducer. However, the major benefit of arrays derives from the ability to be able to drive each element independently and specifically the ability to adjust the precise time that transducer elements produce their acoustic signal.

The simplest ultrasonic transducer "fires" its acoustic signal whenever a voltage is produced by the signal generator/amplifier driving it. However, almost all transducer arrays introduce a variable length delay line to each transducer element so that the instant that each element fires can be precisely controlled. These type of arrays are known as *phased arrays*. This inter-element delay function allows acoustic beams to be steered or focussed in an almost unlimited number of different ways, and provides incredible flexibility in the nature of ultrasonic fields that are produced.

## 6.4.1 Beam steering

Consider initially an array where each element has a delay that is linearly proportional to the distance from the top edge of the array as shown in Figure 6.12. As can be seen, in the very near field the discrete nature of each wavefront can be identified. As these wavefronts propagate forward, diffractive spreading means that they eventually combine to form one continuous wavefront. Note that the wavefront is inclined relative to the surface of the array — the beam has been steered by an angle $\theta$.

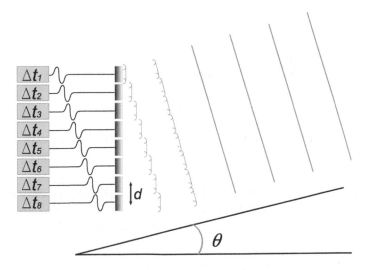

Figure 6.12: Beam steering with an 8-element linear phased array.

Figure 6.12 would tend to suggest that all of the acoustic energy is propagating at one angle only. In practice, two other factors influence the direction at which ultrasound propagates. Firstly, each transducer element will have its own directivity function. Typically elements are rectangular and thus a response of the form (6.21) will probably be appropriate. The directional responses of all elements will contribute to the far-field beam pattern, and thus in addition to the main beam there will also be lower level side lobes. An example of this can be found in Figure 6.13. As was seen in (6.21) the directional response of each element is a function of both the element dimensions and the wave number. Consequently the amplitude and position of side lobes is highly dependant of the geometry of array and its operating frequency.

Figure 6.13 also identifies the second area of concern when looking at the directional response of an array: that of grating lobes. Students of physics will be familiar with diffraction gratings and the equation that describes them, specifically that the angle of intensity maxima is given by

$$n\lambda = d\sin\theta, \text{ where } n \in 1, 2, 3, \dots . \tag{6.33}$$

The elements of an ultrasonic array behave in a very similar manner to the

142

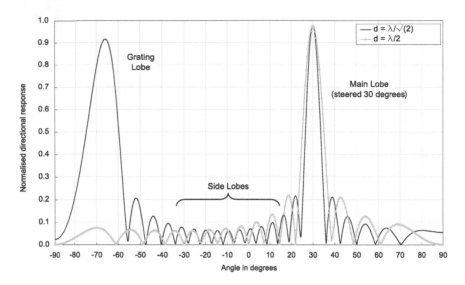

Figure 6.13: Grating and side lobe artefacts on a 16-element array steering a beam by 30°.

slots on a diffraction grating, and thus if the inter-element spacing $d$ becomes too large, grating lobes will be introduced. Wooh and Shi[21] have combined these two effects into a single expression for the directivity of an array in polar coordinates $D(\theta, \gamma)$ such that

$$D(\theta, \gamma) = D_1(\theta, \gamma) \cdot D_2(\theta, \gamma). \tag{6.34}$$

In (6.34), $D_1(\theta, \gamma)$ is the contribution due to the directivity of the individual rectangular elements and is given by

$$D_1(\theta, \gamma) = \left| \mathrm{sinc}\left(\frac{\pi a \sin\theta \cos\gamma}{\lambda}\right) \mathrm{sinc}\left(\frac{\pi L \sin\theta \sin\gamma}{\lambda}\right) \right|, \tag{6.35}$$

whilst the diffraction grating term $D_2(\theta, \gamma)$ is given by

$$D_2(\theta, \gamma) = \left| \frac{\sin\left[N\frac{\pi d}{\lambda}(\sin\theta_S - \sin\theta\cos\gamma)\right]}{N\sin\left[\frac{\pi d}{\lambda}(\sin\theta_S - \sin\theta\cos\gamma)\right]} \right|. \tag{6.36}$$

In both (6.35) and (6.36), $d$ is the inter-element separation, $a$ is the width of an individual element, $L$ is the length of an individual element, $N$ is the number of elements in the phased array group, and $\theta_S$ is the steering angle.

*CAUTIONARY NOTE*: The geometry used to define angles is subtly different between the convention used in this chapter and that of Wooh and Shi. This accounts for the slight difference in form between (6.35) and (6.21).

[21]Wooh S-C. and Shi Y. Three-dimensional beam directivity of phase-steered ultrasound. *J Acoust Soc Am* **1999** 105:3275–3282.

## 6.4.2 Beam focussing

Phased arrays can also be used to focus an ultrasonic beam. This is probably the most widely used of all focussing methods, since this technique is implemented in almost all diagnostic imaging devices. As shown in Figure 6.14, the outermost elements have the shortest delay ($\Delta t_1$) and produce their acoustic signal first. The next pair of elements to fire have delay $\Delta t_2$, with another pair of elements firing shortly after at $\Delta t_3$. The innermost elements are the last to fire after a delay of $\Delta t_4$. As with steered beams, diffractive spreading rapidly ensures a curved wavefront is formed as a composite of the contributions from each individual element.

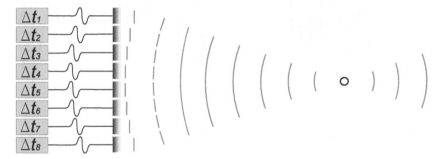

Figure 6.14: Use of an electronic phasing of array elements to produce a curved wavefront.

Since the various delays are not fixed, different focal depths can be obtained by simply changing the relative delays. Almost all diagnostic ultrasound machines exploit this function to allow multiple, user-selectable, focal zones. As drawn in Figure 6.14, the delays are selected so as to achieve a circularly curved wavefront, but alternative shaped profiles (*e.g.*, parabolic or hyperbolic) can by appropriately selected delays. Beam steering is also commonly combined with beam focusing to produce an acoustic beam that is focussed off to one side.

## 6.4.3 Transducer array configurations

Thus far the description of electronically phased arrays has concentrated only on 1-dimensional (1D) linear arrays. However, 1D curved arrays (sometimes referred to as curvilinear arrays) are common. As can be seen from Figure 6.15, whilst a linear array images a parallel sided slice of the subject, a curvilinear array interrogates a sector of a circle, and thus has a wider field of view at depth that it does near the surface. Linear and curvilinear are probably the most widely used diagnostic ultrasound arrays.

The acoustic field that is produced by a 1D phased array is only focussed in the plane parallel to the long axis of the array. However, it may be desirable to prevent the beam from spreading in the elevation direction (*i.e.*, orthogonal to the long axis of the array). This is commonly achieved by placing an acoustic

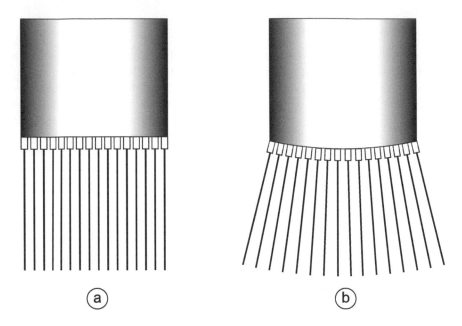

Figure 6.15: Types of 1-dimensional phased array: (a) linear and (b) curvilinear.

lens in front of the array elements to limit the elevational beam spread. This can be seen in Figure 6.16(a). Simple acoustic lenses have a fixed focal length. This can be a limitation, particularly if the primary focus of the phased array is significantly different from that of the front surface lens.

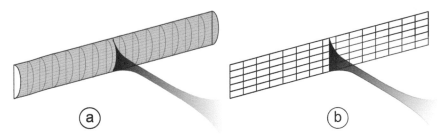

Figure 6.16: Elevational focussing for a phased array: (a) with an acoustic lens and (b) the so-called 1.5D array with additional phased array elements in the elevation direction.

The obvious extension to the simple 1D array is to have a few additional rows of elements so that some form of phased array focussing in the elevation direction is possible. Whilst there may be many array elements ($> 100$) in the principal direction, the array will contain only a limited number of elements (typically 8–12) in the other (elevation) direction. These have become known as 1.5D arrays and an example is shown in Figure 6.16(b). Clearly these devices have additional complexity, but provide a more flexible solution to the problem of elevation focussing. Rather than having a fixed focal length lens, a 1.5D array can have its elevation focal length adapted to suit the focal length in the principal focussing direction. Two-dimensional arrays are the most recent development in ultrasonic array technology and have the advantage of being able to scan a volume (rather than a slice — which is the case with 1D arrays). Furthermore, the ability to be able to independently phase rows and columns allow beam steering in any direction. When coupled with high speed image processing software, 2D arrays enable the clinician to view, rotate, and slice the image in a myriad of different ways to aid diagnosis. A diagram of a 2D array can be found in Figure 6.17(b). The last array type (annular arrays) are

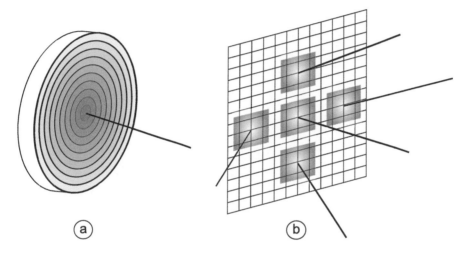

Figure 6.17: Further phased arrays: (a) an annular array and (b) a 2D array.

commonly used in therapeutic ultrasound (particularly HIFU). Annular arrays provide a simple "on-axis" focus, but the focal depth can be varied by altering the phase of the individual annular elements. This is particularly useful when trying to correctly target an ultrasound induced lesion on the region of tissue being treated. Figure 6.17(a) shows a typical annular array configuration.

# 7

# Medical imaging
Knut Matre and Odd Helge Gilja

## 7.1 Standard ultrasonic imaging modes

The use of acoustic waves to obtain 2-dimensional (2D) images of human organs and organ systems was first presented in the 1960s, by applying a single element probe and mechanical movement of this probe, so-called static scanning. Although the images had poor spatial resolution, they provided novel images of the human foetus that caused debate about monitoring of the foetus.

Since then, ultrasonic imaging has undergone an enormous development. Modern scanners for clinical use look similar to the line drawing in Figure 7.1, not much unlike a full-blown personal computer on wheels with several hard drives and integrated peripheral devices, but with ultrasonic probes attached.

This Chapter goes through the different imaging methods and discusses the improvements that have been carried out to make ultrasound one of the most important imaging modalities in medicine. In the early years of ultrasonic imaging, different modalities were obtained using separate instruments, for example, 2D greyscale images and pulsed Doppler measurements required two separate examinations. Modern ultrasound scanners have incorporated all imaging modalities and these are now routinely used together for most clinical examinations.

### 7.1.1 A-mode

Of the single beam display methods, the A-mode (A = amplitude) is the simplest form of displaying the received echo from tissue interfaces. Figure 7.2 shows how the deflection in the $z$-direction of a travelling beam can be proportional to the

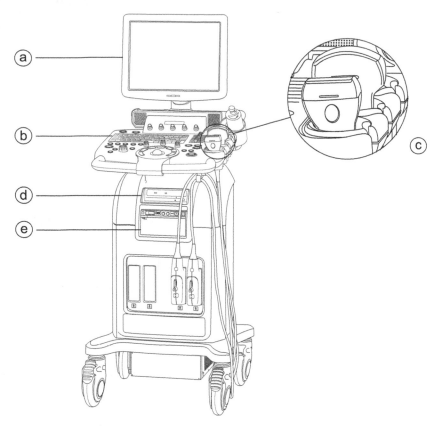

Figure 7.1: Modern ultrasound scanner with a monitor (a), manual controls (b), several probes (c), DVD-RW drive (d), and a printer (e).

amplitude of the received echoes. The $z$-axis can display time from transmitted pulse, but if the ultrasound velocity is known, we can mark the $z$-axis as depth, since

$$d = \frac{ct}{2}, \qquad (7.1)$$

where $d$ is the depth and $c$ is the ultrasound speed. Choosing an average speed of ultrasound for biological tissue makes all dimensional measurements from A-, M-, and B-mode calibrated at this speed. The most commonly used mean ultrasound speed is $1540\,\mathrm{m\,s^{-1}}$. Large deviations from $1540\,\mathrm{m\,s^{-1}}$ will introduce errors in the dimensional measurements.

The A-mode was the first display method used for clinical examinations. Nowadays, it is mainly used for adjustment of the gain, time gain compensation (TGC), also called time varying gain (TVG). It is necessary to obtain an even amplitude registration at all depths to compensate for the attenuation of ultrasound in tissue. The attenuation determines the TGC slope. The A-mode

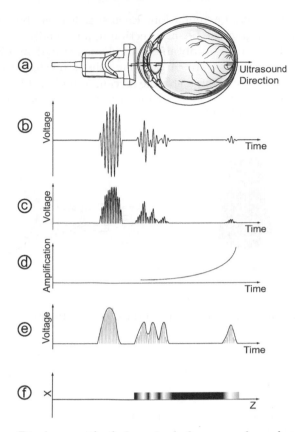

Figure 7.2: Display method for single-beam pulse-echo measurements: (a) ultrasound ray, (b) radio-frequency (RF) data, (c) Hilbert-transformed data, (d) time gain compensation (TGC), (e) A-mode, (f) 1D B-mode.

is a 1-dimensional (1D) imaging system with the amplitude of peaks indicating signal strength.

The maximum distance for imaging, $R_{max}$, is determined by the pulse repetition frequency PRF:

$$R_{max} = \frac{c}{2\text{PRF}}. \tag{7.2}$$

If deeper structures need to be imaged, PRF must be reduced, which follows an adjustment of the depth control.

## 7.1.2 B-mode

If the echoes are displayed as pixels on the screen with the brightness of each pixel corresponding to the strength of the reflected signal, we obtain what is called a B-mode (B = brightness) display. Again, it is necessary to use TGC to compensate for ultrasound attenuation. Each scan line can be considered an A-mode image. By changing the beam direction in a manually controlled manner, a 2D greyscale image can be obtained, as demonstrated in Figure 7.3. The earliest 2D B-mode images were carried out using a mechanical registration of the probe movement. By using a storage screen, a single 2D image was created. This method could not be used on moving objects. The typical PRF for B-mode imaging is between 2 and 4 kHz.

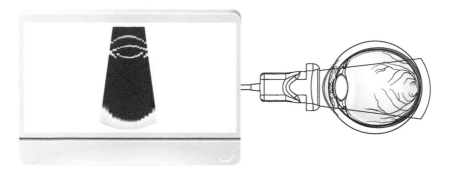

Figure 7.3: Schematic representation of 2D B-mode imaging.

### 7.1.3 M-mode

Displaying the echoes along one beam as a function of time is called M-mode (M = motion), as shown in Figure 7.4. The advantage of this display mode is the excellent time resolution. The sampling frequency can be up to 1 kHz, making it useful to study movement of, for example, heart valve leaflets. The limitation of this display form is that it can only be generated from one beam direction. The so-called curved anatomical M-mode gives the user the opportunity to display the M-mode for a chosen contour in the 2D B-mode image, independent of beam direction.

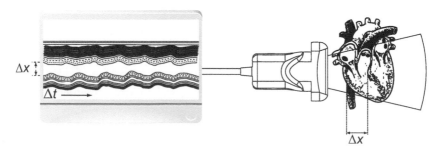

Figure 7.4: Schematic representation of M-mode imaging.

## 7.1.4 Real-time scanning

By moving the scanning beams in a controlled manner and obtaining a frame rate of above approximately 20 frames $s^{-1}$, live images can be obtained. Both revolving and oscillating single element transducers were used in different mechanical sector probes. These were relatively cheap probes, but the disadvantages were, in addition to the vibration from the motor and subsequent wear, a small image view because of the sector format and a low frame rate. The most advanced mechanical sector probe was the annular array, where several separate transducer rings enabled focusing in both lateral and elevation dimensions, thus improving the resolution of 2D images considerably. Nowadays, the annular array is mainly used for endosonographic probes.

Figure 7.5(a) shows the working principle of the linear array developed in the late 1960s. Each scan line with received echoes displayed in B-mode is completed before the next pulse is transmitted from the neighbouring element. Therefore, the scanner must have a number of parallel processing channels matching the number of transmitted beams. Increasing the number of beams increases the lateral resolution of the scanner, but also requires an increasing number of channels, which add to the complexity of the scanner. In contemporary scanners, both transmitting and receiving are carried out using several elements. Electronic focusing is now the standard method for controlling the lateral resolution of the B-mode images.

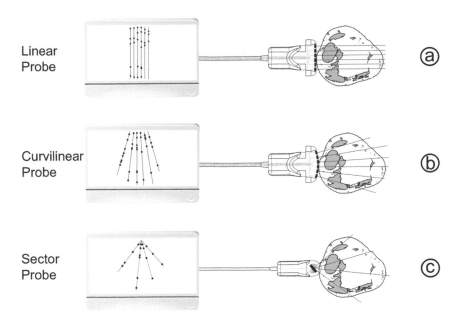

Figure 7.5: Real-time scanning transducer arrays: (a) linear probe, (b) curvilinear probe, (c) sector probe.

Figure 7.5(b) shows the working principle of the curvilinear probe. This is a linear array of elements with a curved probe surface, giving a larger field of view at depths. The disadvantage of this probe is that the beam density is reduced at depths, which causes a reduced lateral resolution.

Figure 7.5(c) shows the working principle of the sector probe. This type of probe is the standard probe for imaging the heart, where small intercostal spaces limit the useful active area, the so-called footprint, of the probe. The sector probe suffers even more from the reduction of beam density at depths. As mentioned earlier, the sector scanning can be obtained using a mechanical movement of a single element transducer. These days, nearly all sector probes are of the phased array type, using electronic steering of the ultrasound beam. Figure 7.6 shows how we can obtain controlled steering of the beam by using six crystal elements during transmission with a short time delay between separate transmission pulses. A typical number of elements is 64. Frame (a) shows all elements activated at the same time, which results in a wavefront transmitted straight forward. Frame (b) shows that a time delay introduced between elements results in a steered wavefront. With the phased array probe, high frame rates can be obtained, typically for the heart 70–90 frames $s^{-1}$ for maximum sector width ($90°$) and 200–300 frames $s^{-1}$ for a smaller sector. Figure 7.7 shows some standard array probes used for diagnosis.

## 7.1.5   Dynamic focus

We can focus a single element transducer by using either a curved element or a lens. This gives a narrow beam (focal zone) at one depth. Figure 7.8 shows the principle of dynamic focus; in this example the received pulse is focussed using six elements. The focal zone can be changed as echoes from increasing depths are expected by altering the time delays. Thus, multiple focal zones can be obtained. Sweeping of the focal zone during receiving is called dynamic focus.

Focussing and steering can be obtained for the linear and curvilinear probes as well as the phased array. An important factor for dynamic focusing is the probe aperture. A large probe aperture (many elements) focusses better than a small probe aperture.

## 7.1.6   Compound scanning

Steering of the ultrasound beam for a linear probe in alternating directions slightly off the normal direction gives better echoes from interfaces that are close to parallel to the beam. This is called compound scanning. Compound scanning improves the images of a circular object like the abdominal aorta in cross-section, since almost the whole circumference will be recorded. The disadvantage of compound scanning is that it only gives half the frame rate of conventional linear scanning.

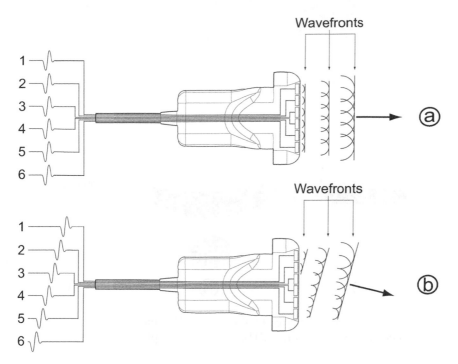

Figure 7.6: Electronic sector scanning using the phased array for beam steering.

## 7.1.7 Curved anatomical M-mode

Instead of displaying the M-mode of echoes along one beam, ultrasound scanners have the possibility to draw a curved line in a stored 2D B-mode image and to display the echoes from this trajectory. This is called the anatomical M-mode. Figure 7.9 shows how the curved anatomical M-mode from the ventricular septum of the heart is perceived.

## 7.1.8 Resolution of a B-mode image

In B-mode imaging, the image resolution is defined as the minimum distance between two small objects that appear as two objects on the screen. Moving the two objects any closer will result in only one echo on the display. The resolution is usually different in different directions, labelled as axial resolution along the beam, also called radial resolution for sector image formats, lateral resolution across the beam (in-plane), and elevational or azimuthal for slice thickness (*cf.* Figure 7.10). The latter is normally controlled by a fixed lens giving some focusing of the slice thickness over a relatively large depth range. This can be improved by using a 1.5D matrix probe employing five arrays of crystals and with the use of time delays, as shown in Figure 6.16.

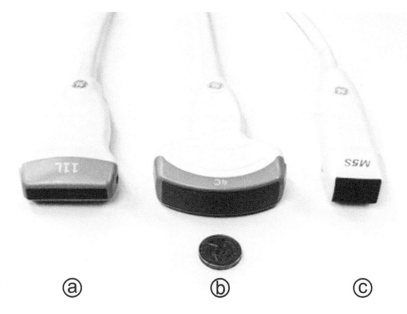

Figure 7.7: Standard array probes: 11-MHz linear array (a), 4-MHz curvilinear array (b), and 5-MHz, 1.5D matrix phased array (c).

The axial resolution $\mathcal{R}$ is defined by

$$\mathcal{R} = \frac{c\tau}{2} = \frac{c}{2B}, \tag{7.3}$$

where $\tau$ is the pulse length and $B$ is the pulse bandwidth. Thus, the axial resolution is proportional to the pulse length. The improvement of images in the 1980s and 1990s was a result of the development of wide-band transducers that were able to transmit short pulses.

Across the beam, the in-plane lateral resolution $\mathcal{L}$ is mainly determined by the size of the probe. For a spherical single element, it is defined by

$$\mathcal{L} = \frac{\lambda F}{D}, \tag{7.4}$$

where $D$ is the transducer aperture, $F$ is the focal depth, and $\lambda$ is the wavelength. Thus, the resolution improves with larger aperture.

## 7.1.9 Factors affecting image quality

Table 7.1 shows the most important factors determining image quality; some of these can be controlled by the operator. A compromise between resolution and penetration has to be determined for each application and also for each patient. It is necessary to obtain sufficient penetration, which can primarily be obtained by reducing the transmit frequency. Table 7.2 gives frequencies and penetration

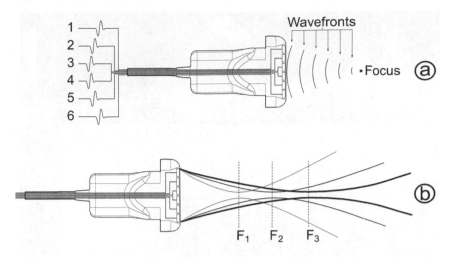

Figure 7.8: Principle of dynamic focus: (a) A spherical wavefront is sensed; the signals from the central elements are time-delayed before adding all signals from all elements, so that the focus can be changed by changing time delays. (b) Multiple foci, here indicated by focal zones F1–F3.

depths for routine ultrasound imaging of specific organs. Examples of B-mode

| Factor | Ameliorated by |
|---|---|
| Resolution | Higher frequency |
| | Dynamic focus |
| | Larger transducer aperture |
| Penetration | Lower frequency |
| | Higher intensity |
| Frame rate | Lower penetration depth |
| | Smaller sector angle |
| Dynamic range | Higher number of greyscale levels |
| | Lower noise level |
| | Harmonic imaging |
| Artefact rejection | Harmonic imaging |
| | Smaller side lobes |

Table 7.1: Quality factors for B-mode imaging.

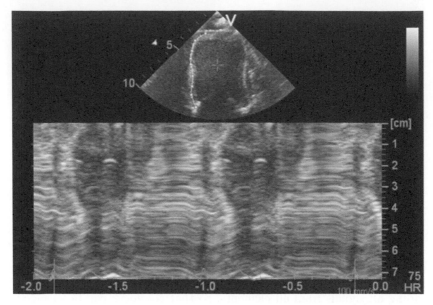

Figure 7.9: Anatomical M-mode of the ventricular septum.

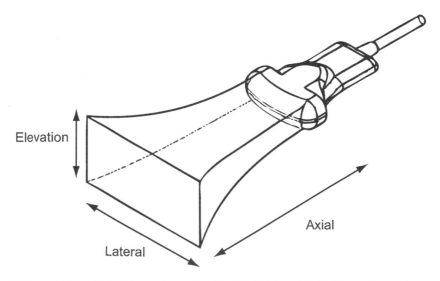

Figure 7.10: Axial, lateral, and azimuthal (elevation) dimensions of an array probe.

images are shown in Figures 7.11–7.14.

Today, higher frequencies are used for the same application than five to ten years ago. Improved instruments with better penetration have lead to an increase in standard probe frequency. As an example, the standard adult cardiac probes have shifted from 2.5–3.3 up to 4–5 MHz. Also, for wide-band probes, enabling the adjustment of the frequency during scanning has made it possible to optimise the acquisition for a specific patient condition. An adult phased array probe of 4 MHz can be typically used at 3.3–5 MHz. These frequencies are valid for fundamental imaging where the same frequency (band) is used for both transmission and receiving. Other frequencies are used for harmonic imaging, as discussed in the next Section.

The ability to display both strong and weak echoes in the same image, *i.e.*, a high dynamic range, is an important factor for image quality. Sometimes the pathology appears in the form of strong echoes, as with a calcified heart valve, but sometimes as a weak echo, as with a small tumour.

The frame rate itself might be considered a quality factor for moving objects, such as the myocardium and heart valves. Here, the user needs to adjust depth and sector angle to obtain sufficient frame rate.

## 7.1.10 Harmonic imaging

The method of harmonic imaging was first developed for contrast microbubbles. Tissue harmonic imaging (THI) exploits the gradual generation of higher frequency harmonics as the ultrasound pulse travels through biological tissue. Several advantages are obtained by using the second harmonic components generated by nonlinear processes. Higher-than-second harmonics are also received,

| Fundamental (MHz) | Harmonic (MHz) | Penetration (cm) | Organs |
|---|---|---|---|
| 2–3 | 1.6–2.8 | 30 | Deep abdomen |
| 4–5 | 3.6–4.6 | 20 | Adult heart |
| 6–7 | 4.2–6.6 | 10 | Superficial abdomen |
| 8–10 | 5.6–9.4 | 4 | Peripheral vessels / Paediatric heart / Endosonographic |
| 12–15 | 10–14 | 3 | Skin / Eye / Mammae / Thyroid / Intravascular / Endosonographic |
| 20–50 | — | 1 | Skin / Eye |

Table 7.2: Typical B-mode applications.

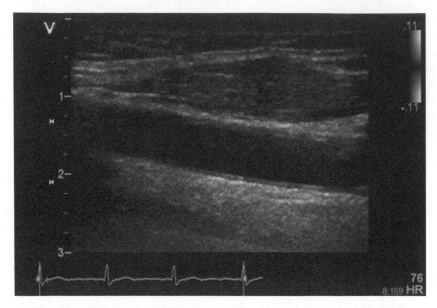

Figure 7.11: B-mode image of the common carotid artery using an 11-MHz linear array.

but with an increased reduction in signal amplitude. So far, only the second harmonic components have been used for clinical B-mode imaging.

An artefact is a displayed echo not generated by an anatomic structure but by the imaging system itself. Near-field artefacts, like reverberation, are suppressed when using harmonic imaging, because there are only fundamental frequencies at the transducer face and the nonlinear processes are not significant until after some distance from the transducer. The near field also produces image clutter that is difficult to interpret, as well as echoes from side lobes. THI reduces image clutter and noise and therefore improves the signal-to-noise ratio of the image.[1] THI requires a receiver system with a high dynamic range and sensitivity. It was first believed that this method could only be used for ultrasound contrast imaging. Improvements in system design have made THI a method used in routine B-mode imaging, too. Figure 7.15 shows typical images from the abdomen using fundamental and harmonic imaging.

Harmonic images contain less noise and as a result more details can be seen. For higher frequencies with lower penetration, less higher harmonics are produced and the effect of using harmonic imaging is reduced.

---

[1]Tranquart F, Grenier N, Eder V, Pourcelot L. Clinical use of ultrasound tissue harmonic imaging. *Ultrasound Med Biol* **1999** 25:889–894.

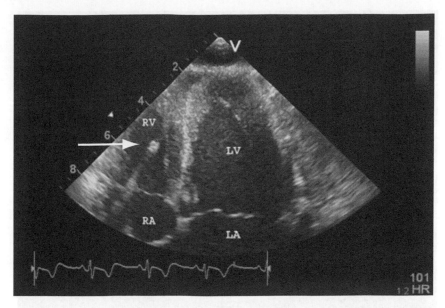

Figure 7.12: Apical four-chamber B-mode image of the heart using a 5-MHz, 1.5D matrix phased array probe. LA = left atrium, LV = left ventricle, RA = right atrium, RV = right ventricle. Note the strong reflection from a Swan–Ganz catheter in the RV for cardiac output measurements (arrow).

## 7.1.11  3D/4D B-mode methods

3D and later 4D (*i.e.*, a time sequence of 3D) B-mode imaging of the heart was introduced using mechanical movement of the imaging plane. These early systems were not real-time; acquisition of images at a specific point in the cardiac cycle was obtained from a number of heart beats, making the method difficult to use. Limitations on image resolution and computer power of workstations led to time-consuming rendering processes and made these methods a research tool only. Less critical objects with respect to movement like the foetus and abdominal organs can be imaged with mechanical steering of the imaging plane. One of the most popular clinical systems uses a curvilinear array with a mechanical movement. The disadvantage of this method is a low frame rate. Full (2D) matrix probes have been developed and are getting increased clinical attention for cardiac imaging.

In addition to the transcutaneous 3D/4D methods, 3D transoesophageal probes for the heart have gained popularity for imaging cardiac valves. On the abdomen, transducers with a position sensor are used in addition to the electronic/mechanical probes discussed above. Here a small position sensor is attached to a standard abdominal probe. Figure 7.16 shows a 3D image of the pancreatic duct obtained with a position sensor.

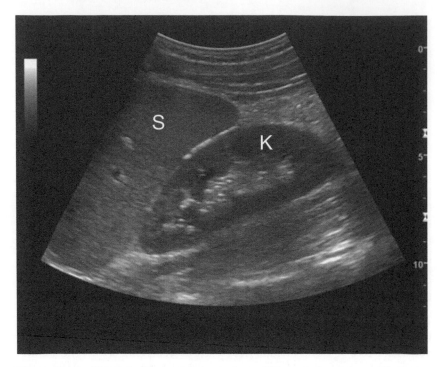

Figure 7.13: B-mode image of the kidney (K) and the spleen (S) using a 4-MHz curvilinear abdominal probe.

For the heart, several arrangements for moving the imaging plane have been published, but the full matrix probe is dominant in cardiac ultrasound, which avoids the artefacts introduced by gated acquisition.[2] An example of a 3D image of the heart is shown in Figure 7.17. The frame rate is still an important limitation of this method. Typical frame rates are 16–25 frames $s^{-1}$.

## 7.2 Doppler methods

Methods based on the registration of the Doppler shift caused by moving reflectors have become an increasingly important part of routine ultrasound examinations. We can divide these methods into single-beam Doppler methods and 2D Doppler methods.

### 7.2.1 Single-beam Doppler methods

Figure 7.18 shows the different methods where the Doppler shift is registered from one Doppler beam or a limited number of beams transmitted in the same

---

[2]McCulloch ML, Little SH. Imaging methodology and protocols for three-dimensional echocardiography. *Curr Opin Cardiol* **2009** 24:395–401.

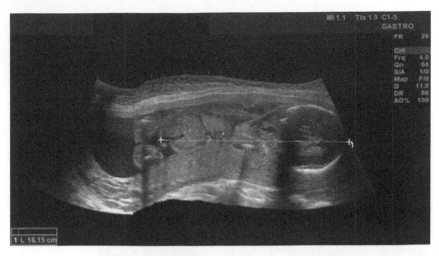

Figure 7.14: Using the panoramic function, the whole foetus can be visualised with a 4-MHz probe.

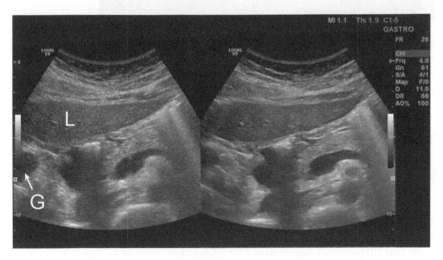

Figure 7.15: Fundamental (*left*) and second harmonic (*right*) imaging of the liver (L) and gallbladder (G), using a broadband 1–5 MHz probe.

direction. They are all based on the Doppler equation, which for a stationary transmitter/receiver and a moving reflector, moving in the B-mode plane of observation at a velocity $v$ and a direction $\theta$ relative to the transducer axis, is found by combining (4.66) and (4.68). The resulting Doppler shift is defined by

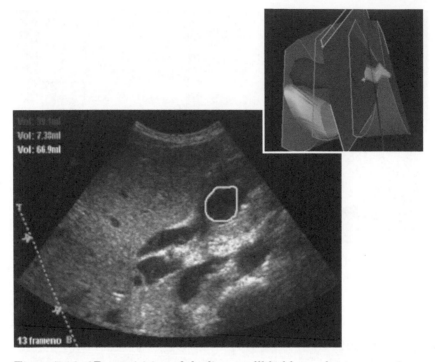

Figure 7.16: 3D acquisition of the liver, gallbladder and pancreas using a magnetic position sensor: Delineation of a pancreatic cyst (*lower panel*) and a 3D object reconstruction by manual contour detection (*upper panel*).

the difference in observed frequency and the transmitted frequency:

$$f_D = \frac{2\frac{v}{c}\cos\theta}{1 - \frac{v}{c}\cos\theta} f_0, \tag{7.5}$$

where $f_D$ is the Doppler shift, $f_0$ is the transmitted frequency, $v$ is the magnitude of the velocity of the blood, $c$ is the speed of sound in blood, and $\theta$ is the angle between the ultrasound beam and the blood flow direction. Since $v \ll c$,

$$f_D = 2f_0 \frac{v}{c} \cos\theta. \tag{7.6}$$

Thus, the Doppler shift will be positive if the movement is towards the probe and negative if away from the probe. Note that in order for the blood velocity to be measured, the angle must be known or be small. Only the velocity component along the axial beam is measured. The Doppler-shifted signal is demodulated. This so-called Doppler signal is in the audible range and can be listen to via a loudspeaker. Figure 7.19 shows the Doppler shift as a function of transmitted frequency and velocity. The Doppler shift is proportional to the transmitted

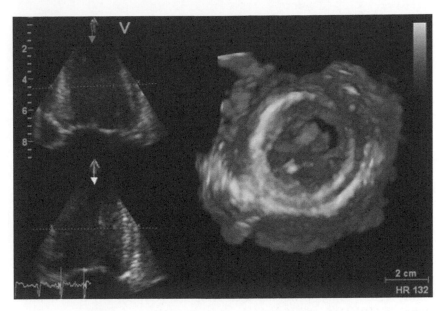

Figure 7.17: 3D image of the heart using a 3-MHz matrix probe: two perpendicular planes of the left ventricle (*left*) and a volume-rendered 3D short axis of the left ventricle looking towards the mitral valve (*right*).

frequency; a two-fold increase in transmitted frequency gives twice the Doppler shift.

We often use a lower transmitting frequency for the Doppler pulse than for B-mode pulse, since reflections from blood are weaker than reflections from tissue interfaces.

## 7.2.2 Continuous wave Doppler

As indicated in Figure 7.18(a), in the continuous wave Doppler (CWD) mode the transmit element sends out a continuous wave of ultrasound and a separate crystal is used for receiving the reflected ultrasound from both stationary and moving reflectors. If we filter out the ultrasound from stationary targets, which would have a frequency equal or very close to the transmitting frequency, we can detect the Doppler-shifted frequencies. The CWD is sensitive to Doppler-shifted reflections only from the hatched area in Figure 7.18(a).

## 7.2.3 Pulsed wave Doppler

The pulsed wave Doppler (PWD) mode is using the same crystal for transmitting and receiving ultrasound, as shown in Figure 7.18(b), similar to B-mode imaging. A short pulse of typically 6–12 cycles is transmitted with a typical

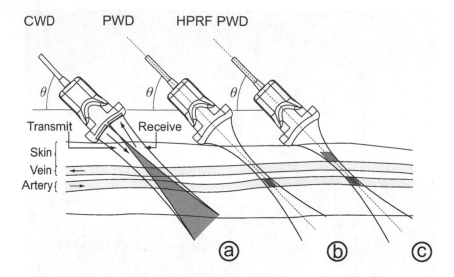

Figure 7.18: (a) Continuous wave Doppler (CWD), (b) pulsed wave Doppler (PWD), and (c) high pulse repetition frequency Doppler (HPRF).

PRF of 5–15 kHz. The receiver is opened only for a short time window, called the range cell or sample volume. This increases the axial resolution by ignoring the Doppler shift until a set time after transmission. This time is seen as depth in centimetres and not as time in microseconds by the observer. The increase in axial resolution is an advantage of this method, but the trade-off is that with a short ultrasound pulse, the frequency cannot be determined accurately. So there is a compromise between the accuracy of Doppler-shift measurements and depth resolution (axial resolution).

An additional problem with PWD is aliasing, which is introduced by the PRF. Here, the Doppler frequency is sampled with the PRF. According to Shannon's sampling theorem the PRF should be at least twice the highest frequency in the signal. The highest Doppler shift that can be sampled without aliasing is thus half the PRF, the Nyquist frequency:

$$f_{D,max} = \frac{PRF}{2}. \tag{7.7}$$

The corresponding blood velocity is called the Nyquist velocity. Since the PRF is related to the penetration depth, the maximum velocity that can be measured is also related to the depth range. Combining (7.2), (7.6), and (7.7) gives the maximum unambiguous blood velocity as a function of the maximum distance imaged:

$$v_{max} = \frac{c^2}{8 f_0 R_{max}}. \tag{7.8}$$

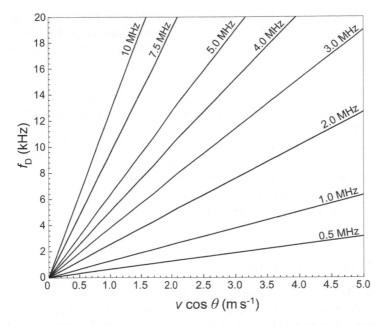

Figure 7.19: Doppler shift versus lateral velocity for three different transmitting frequencies.

For example, at a distance of 10 cm, a 5-MHz Doppler system can detect velocities up to approximately $0.5\,\mathrm{m\,s^{-1}}$ without aliasing.

## 7.2.4 High pulse repetition frequency Doppler

A method which enables the use of PWD for higher velocities is the high pulse repetition frequency (HPRF) Doppler. Here, the next pulse is transmitted before the echoes from the first sample volume have been registered. Thus, the HPRF Doppler has several sample volumes along the beam and a depth ambiguity is introduced. If the examiner is careful with the placement of the different sample volumes, with only one sample volume covering a blood vessel, high velocities can be measured in the pulsed mode. Figure 7.18(c) shows HPRF PWD with only sampling in the artery. The HPRF Doppler is a standard mode for most modern scanners. For many of these, this mode is automatically activated when the user is adjusting the velocity scale outside the range giving no aliasing. Using only one sample volume along the beam is called low pulse repetition frequency (LPRF) Doppler.

## 7.2.5 Directivity and spectral analysis

The RF-amplified Doppler-shifted signal is fed to a quadrature demodulator, which multiplies the signal with both the cosine and sine of the transmitted frequency. This gives both the magnitude and sign of the Doppler signal. An important property of CWD, PWD, and colour Doppler is the ability to detect the direction of the velocity, towards (positive) and away from the probe (negative).

The Doppler signal always contains a band of frequencies, because the reflections from blood cells across the vessel have different velocities, but also because of the limitations of the method. After demodulation, the signal is digitised and undergoes a Fourier-spectral analysis. This signal is displayed as a time curve with a velocity distribution for typically every 5 ms. Figure 7.20 shows a spectral curve from the normal carotid artery with a narrow band of velocities in the acceleration of blood during systole, *i.e.*, a flat velocity profile, and a wider band of velocities during diastole, *i.e.*, a parabolic velocity profile. Thus, the spectral analysis gives information of blood flow quality as well as velocity information.

## 7.2.6 Duplex scanning

Combining PWD and B-mode imaging, as in Figure 7.20, is called duplex scanning. For the time intervals of B-mode measurements, the Doppler signal is substituted with a synthetic signal from a missing signal estimator to produce a continuous Doppler signal.[3] A lower frame rate for the B-mode image results

---

[3]Kristoffersen K, Angelen BAJ. A time-shared ultrasound Doppler measurement and 2-D imaging system. *IEEE Trans Biomed Eng* **1988** 35:285–295.

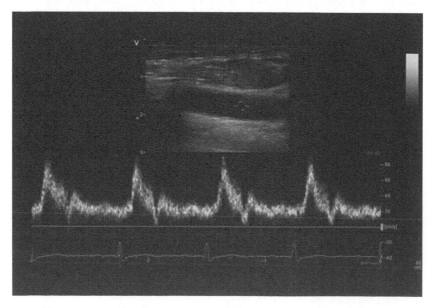

Figure 7.20: Duplex scanning of the common carotid artery.

for this combination. Most ultrasound scanners have the possibility to freeze the B-mode image and update it with a low or high frame rate, gradually decreasing the quality of the Doppler acquisition as a result. The duplex scanning mode makes Doppler methods easier to use and improves diagnosis. However, one should keep in mind that the ultrasound beam should preferably be perpendicular to tissue interfaces and parallel to blood vessels. Figure 7.21 shows an example of duplex scanning in the abdomen whilst recording high velocities with HPRF Doppler.

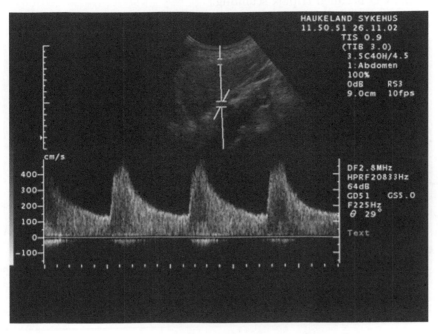

Figure 7.21: Measuring high velocities in the arteria mesenterica superior, whilst using duplex scanning with HPRF Doppler.

### 7.2.7  Colour Doppler

Using many sample volumes along the beam requires many Doppler scan line processor channels in parallel. Sweeping this beam with multiple sample volumes give velocity information in a 2D area. The mean velocity is then converted to a colour code for online interpretation. This method is called colour Doppler (CD) or colour flow mapping (CFM). Usually the resolution for colour Doppler is less than for B-mode, because a large beam width and sample volume are required for getting sufficient echoes from moving blood, resulting in large colour picture elements. An important limitation of the colour Doppler method is that it goes into aliasing often at normal arterial velocities, because it is a pulsed Doppler system. The standard colour coding is here red for velocities towards the probe, blue for velocities away from the probe, and yellow or green for velocities above the Nyquist limit. It is important to note that colour Doppler only measures velocities along the beam and can not detect velocities across the beam. Figure 7.22 shows a colour Doppler recording of the common carotid artery. The colour Doppler mode is part of most contemporary ultrasound examinations. It is used to visualise large and small vessels and to identify hypo-echoic regions in the B-mode image. These could be different fluid lumina like an artery, vein, or a cyst. Combining B-mode imaging with both PWD and CD is called triplex scanning.

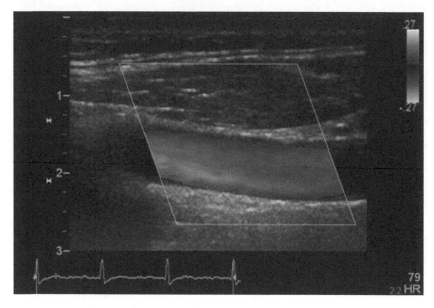

Figure 7.22: Colour Doppler of the common carotid artery, here shown in greyscale. The virtual box in which the velocities are displayed in colour is user-defined.

## 7.2.8 Power Doppler

An alternative colour coding is called power Doppler. Instead of the velocity, here the signal power is displayed as one colour (yellow) with a longer time constant introduced. This method is useful for detecting small vessels and also for semi-quantitative measurement of blood perfusion and measurements of the vessel lumen.

### 7.2.9   Blood flow measurement

It is possible to measure blood flow by combining PWD and B-mode or power Doppler. Blood flow is given by:

$$Q = Av \cos \theta, \tag{7.9}$$

where $Q$ is blood flow (volume per duration), $v$ is the velocity averaged in space and time, and $A$ is the cross-sectional area. Blood flow is normally estimated from the velocity in PWD mode with a large sample volume to cover both the highest velocity at the centre of the vessel and the lowest velocity close to the vessel wall. This mean velocity is then time-averaged over one or multiple heart cycles and multiplied with the cross-sectional area. The vessel cross-section can be measured from the B-mode image or from a power Doppler image. Figure 7.23 shows an example of flow measurement in the normal common carotid artery. Despite the limited precision of the method, in normal straight sections of an artery, good estimates can be obtained. There are few alternatives for completely non-invasive blood flow measurements; magnetic resonance imaging (MRI) is the only relevant option. The cardiac output of the heart can be measured in a similar manner. The left ventricular outflow tract is close to cylindrical and thus both the cross-section and velocity distribution are quite symmetrical, even though some skewness in the velocity profile has been shown both in the left ventricular outflow tract and the ascending aorta.[4,5]

## 7.3   Special techniques

There are several special ultrasound imaging techniques, which are performed by different medical specialities, for example the special techniques for investigating the gastrointestinal tract carried out by a gastroenterologist.[6]

### 7.3.1   Endosonographic methods

If the ultrasound array is small enough to be swallowed, the penetration depth can be lowered and the frequency can be increased to obtain higher resolution. Transoesophageal probes are used both in cardiology and in gastroenterology. The latter probe usually also includes optical imaging. Other endosonographic methods in use are transvaginal, transrectal, and intraluminal imaging of arteries such as coronary arteries.

---

[4]Segadal L, Matre K. Blood velocity distribution in the human ascending aorta. *Circulation* **1987** 76:90–100.

[5]Zhou YQ, Færestrand S, Matre K, Birkeland S. Velocity distributions in the left ventricular outflow tract and the aortic anulus measured with Doppler colour flow mapping in normal subjects. *Eur Heart J* **1993** 14:1179–1188.

[6]Ødegaard S, Hausken T, Gilja OH. *Basic and New Aspects of Gastrointestinal Ultrasonography.* Singapore: World Scientific Publishing **2005**.

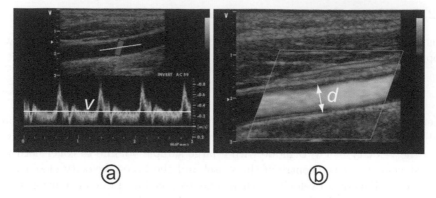

Figure 7.23: Flow measurements in the common carotid artery by combining velocity and cross-sectional area measurement: (a) A large sample volume (dark grey) covering most of the vessel gives a peak systolic velocity of $0.8\,\mathrm{m\,s^{-1}}$. The user-adjusted central line in the vessel indicates the flow direction for automatic velocity scale adjustment. The white line in the velocity spectrum indicates the time-averaged velocity, here $0.3\,\mathrm{m\,s^{-1}}$. (b) The cross-sectional area is obtained from the vessel diameter $d$ in the power Doppler image. Here, $Q = 424\,\mathrm{ml\,min^{-1}}$.

### 7.3.2 Ultrasound-guided biopsy

Most biopsies in the abdomen are currently performed using ultrasound guidance. Special probes with an open slit for the biopsy needle or attachments to a standard ultrasound probe are used. The biopsy needle is visualised in the B-mode image for safer and more accurate biopsies.

### 7.3.3 Tissue Doppler imaging (TDI)

If, by filtering, the Doppler shift from blood is removed and the Doppler shift from moving tissue is detected, the velocity of the myocardium can be measured.[7] Furthermore, the velocities along the beam placed in the myocardial wall can be used to estimate the strain and rate of strain.[8] This has become useful for detecting myocardial dysfunction, e.g., myocardial ischemia, and can also be used for other contracting muscles, such as the stomach.[9] An alternative is to use so-called speckle tracking.[10] This can also be used for non-contracting

---

[7]McDicken WN, Sutherland GR, Moran CM, Gordon LN. Colour Doppler velocity imaging of the myocardium. *Ultrasound Med Biol* **1992** 18:651–654.

[8]Heimdal A, Støylen A, Torp H, Skjærpe T. Real-time strain rate imaging of the left ventricle by ultrasound. *J Am Soc Echocardiogr* **1998** 11:1013–1019.

[9]Gilja OH, Heimdal A, Hausken T, Gregersen H, Matre K, Berstad A, Ødegaard S. Strain during gastric contractions can be measured using Doppler ultrasonography. *Ultrasound Med Biol* **2002** 28:1457–1465.

[10]Leitman M, Lysyansky P, Sidenko S, Shir V, Peleg E, Binenbaum M, Kaluski E, Krakover R, Vered Z. Two-dimensional strain: a novel software for real-time quantitative echocardio-

tissue by inducing deformation of the tissue by cyclic movement of the probe, known as elastography.[11]

## 7.4 Artefacts

An important challenge in the clinical use of ultrasound is to identify and reduce the different artefacts that frequently appear during scanning. In this section, only the most common types of artefacts are mentioned. Many ultrasonic artefacts have been reduced in modern scanners, owing to improvements in signal processing.

### 7.4.1 Attenuation

When the ultrasound pulse hits gas, bone, or another material that has an acoustic impedance very different from biological soft tissue, we get a strong echo from this transition. The ultrasound is attenuated too, so that a shadow appears behind the structure. This is called an attenuation or shadow artefact. An example is shown in Figure 7.24, where a stone is present in the gallbladder. Similar attenuation artefact appear for most artificial materials, like the catheter in the right ventricle shown in Figure 7.12, or any implant.

### 7.4.2 Reverberation

The other most common artefact is reverberation. When the reflected pulse hits a strong reflecting interface like the skin or the probe itself, the ultrasound pulse makes another passage and a deeper situated echo that does not correspond to any anatomical interface appears on the screen. Reverberation can be observed to some extend in most images. Figure 7.24 shows reverberation where the ultrasound enters the gallbladder (arrowhead). This recording has both reverberation and shadowing artefacts. Reverberation can be removed by using second harmonic imaging and by reducing the power level.

### 7.4.3 Mirror artefact

The mirror artefact is similar to reverberation. Here, the artefact is generated from multiple beam reflections between a organ with strong interface echoes and a strong reflector, like the diaphragm. The artefact appears as a virtual object, as if there were a mirror. The reflector can be at an angle to the ultrasound beam.

graphic assessment of myocardial function. *J Am Soc Echocardiogr* **2004** 17:1021–1029.

[11]Ophir J, Garra B, Kallel F, Konofagou E, Krouskop T, Righetti R, Varghese T. Elastographic imaging. *Ultrasound Med Biol* **2000** 26:S23–S29.

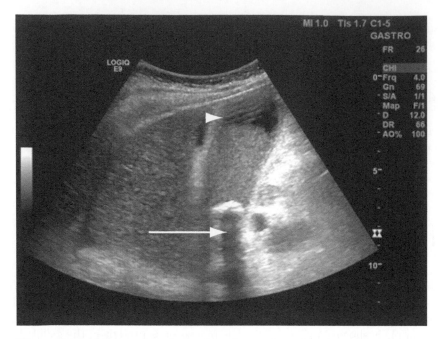

Figure 7.24: Attenuation (*arrow*) and reverberation (*arrowhead*) arte-facts in an image of the gallbladder.

## 7.4.4 Side lobes

In B-mode imaging, side lobes of the ultrasound beam can cause anatomy out-side the main beam to be mapped into the main beam. Side lobe artefacts may appear in the Doppler mode as well. The side lobe could hit a blood vessel while the main beam is outside the vessel. With colour Doppler, the colour presentation of the blood velocity could be misplaced in the B-mode image.

## 7.4.5 Other artefacts

Other artefacts include geometric distortion caused by variations in sound speed that refract the beam. Wave front aberration caused by irregular variation in sound speed could occur. For a more comprehensive discussion of all artefacts, we refer to some recent publications.[12,13]

---

[12]Rubens DJ, Bhatt S, Nedelka S, Cullinan J. Doppler artifacts and pitfalls. *Radiol Clin North Am* **2006** 44:805–835.

[13]Feldman MK, Katyal S, Blackwood MS. US artifacts. *Radiographics* **2009** 29:1179–1189.

# 7.5 Biological effects of ultrasound and safety regulations

Since the start of the clinical use of ultrasound, there have been numerous studies on the effects of ultrasound on macromolecules, cells, animals, and patients subjected to ultrasound exposure. Early studies on chromosome aberrations of white blood cells subjected to a low-intensity ultrasound foetal Doppler instrument showed alarming results.[14] However, these results have not been verified by others[15] and some of the findings have been withdrawn.[16]

High-intensity ultrasound has been used to obtain an understanding of the mechanisms of the interaction between ultrasound and biological tissue. Data have been obtained to find the potential roles of sound speed, intensity, frequency, and attenuation in the occurrence of a raised temperature, cavitation, streaming and radiation. Before 1993, $100 \, \mathrm{mW \, cm^{-2}}$ was used as an upper intensity limit for human exposure. This limit was determined by the spatial average/temporal average intensity (SATA). This limit had been based on the finding that tissue with high perfusion, $e.g.$, heart and kidneys, does not experience a significant temperature increase at ultrasound intensities below $100 \, \mathrm{mW \, cm^{-2}}$ SATA. However, new equipment using dynamic focusing has lead to much higher local intensity peaks. The spatial peak/temporal average (SPTA) intensity is a better measure for the local intensity. In 1993, the United States Food and Drug Administration introduced an overall intensity limit of $720 \, \mathrm{mW \, cm^{-2}}$ SPTA. For ophthalmic applications the corresponding limit was set to $50 \, \mathrm{mW \, cm^{-2}}$ SPTA and for foetal heart rate monitors to $20 \, \mathrm{mW \, cm^{-2}}$ SPTA. It was also decided that the exposure level should be communicated to the user by means of indices on the scanner display, giving information on potential hazard. Because adverse effects of clinical ultrasound have been related to temperature elevation and inertial cavitation, modern ultrasound equipment shows two related indices: the thermal index and the mechanical index.[17] These indices change with ultrasound mode and acquisition settings.

## 7.5.1 Thermal index

The thermal index (TI) is a measure of the temperature rise in the tissue during exposure. It is defined by

$$\mathrm{TI} = \frac{W}{W_{\mathrm{deg}}}, \qquad (7.10)$$

---

[14]Macintosh IJ, Davey DA. Chromosome aberrations induced by an ultrasonic fetal pulse detector. *Br Med J* **1970** 4:92–93.

[15]Lucas M, Mullarkey M, Abdulla U. Study of chromosomes in the newborn after ultrasonic fetal heart monitoring in labour. *Br Med J* **1972** 3:795–796

[16]Macintosh IJ, Davey DA. Relationship between intensity of ultrasound and induction of chromosome aberrations. *Br J Radiol* **1972** 45:320–327.

[17]Abbott JG. Rationale and derivation of MI and TI: a review. *Ultrasound Med Biol* **1999** 25:431–441.

where $W$ is the transmitted power and $W_{\mathrm{deg}}$ is the estimated power needed to raise the tissue temperature 1°C. It should be noted that TI does not indicate the actual temperature rise in the tissue. Since different tissues give different $W_{\mathrm{deg}}$, several thermal indices have been introduced. Three commonly used thermal indices are the thermal index of soft tissue (TIS), the thermal index of bone (TIB) and the thermal index of cranial bone (TIC). The relevance of these has been widely discussed.[18] Based on thermal indices, limitations of exposure times have been recommended. For example, at TI = 2.0 the examination should not exceed 60 min, whereas at TI = 3.0 duration should be limited to 15 min.

## 7.5.2   Mechanical index

The mechanical index (MI) gives an indication of mechanical damage of tissue due to inertial cavitation. It is defined by

$$\mathrm{MI} = \frac{p^-}{\sqrt{f_{\mathrm{c}}}}, \tag{7.11}$$

where $p^-$ is the maximum value of peak negative pressure anywhere in the ultrasound field, measured in water but reduced by an attenuation factor equal to that which would be produced by a medium having an attenuation coefficient of $0.3\,\mathrm{dB\,cm^{-1}\,MHz^{-1}}$, normalised by 1 MPa, and $f_{\mathrm{c}}$ is the centre frequency of the ultrasound normalised by 1 MHz. For MI < 0.3, the acoustic amplitude is considered low. For 0.3 < MI < 0.7, there is a possibility of minor damage to neonatal lung or intestine.[19] These are considered moderate acoustic amplitudes. For MI > 0.7, there is a risk of cavitation if an ultrasound contrast agent containing gas microspheres is being used, and there is a theoretical risk of cavitation without the presence of ultrasound contrast agents.[20] The risk increases with MI values above this threshold. These are considered high acoustic amplitudes. On commercial scanners, the MI has been limited to 1.9 for medical imaging.[21]

Most obstetric investigations are carried out with both TI and MI lower than 1.0; higher values occurs only during short periods of Doppler application.[22,23]

As opposed to the TI, there is only one, primitive, tissue model used for the calculation of the MI. Therefore, the accuracy and general applicability have

[18]Duck FA. Medical and non-medical protection standards for ultrasound and infrasound. *Prog Biophys Mol Biol* **2007** 93:176–191.

[19]British Medical Ultrasound Society. *Guidelines for the safe use of diagnostic ultrasound equipment* **2000**.

[20]ter Haar G. Safety and bio-effects of ultrasound contrast agents. *Med Biol Eng Comput* **2009** 47:893–900.

[21]Voigt JU. Ultrasound molecular imaging. *Methods* **2009** 48:92–97.

[22]Deane C, Lees C. Doppler obstetric ultrasound: a graphical display of temporal changes in safety indices. *Ultrasound Obstet Gynecol* **2000** 15:418–423.

[23]Sheiner E, Freeman J, Abramowicz JS. Acoustic output as measured by mechanical and thermal indices during routine obstetric ultrasound examinations. *J Ultrasound Med* **2005** 24:1665–1670.

been under discussion, especially when using contrast agents. An alternative mechanical index will have to be developed.

### 7.5.3 Clinical studies

Of special interest for the safety of clinical ultrasound are the many follow-up studies of school children exposed to ultrasound *in utero*. There has been no association between ultrasound and malignancies,[24] but an increase in left-handedness among male subjects has been found.[25,26] This association is poorly understood. A recent meta-analysis of all available studies on the safety of ultrasonography in pregnancy concluded that exposure to diagnostic ultrasonography appears to be safe.[27]

### 7.5.4 Concluding remarks on biological effects

Even though there are many exposure conditions during ultrasound examination where the risk is clearly negligible, other exposures, including Doppler modes, can lead to temperature rises and could be harmful. The user of ultrasound equipment should be aware of the possible harmful effects, and use the equipment according to the recommendations from national and international ultrasound societies. There are some differences in the recommendations of ultrasound exposure between different countries. The similarities have been summarised in a recent review.[28]

Of serious concern is the non-medical use of foetal ultrasound.[29] Most health experts and clinicians agree that such use should be avoided and ultrasound should be considered a medical tool for providing clinical information. We are obliged to closely monitor the reports on biological effects, especially when considering the technical improvements and modifications of ultrasound instrumentation, because these often lead to higher ultrasound exposure. An example is improved focusing techniques. A good principle is that of ALARA: As Low As Reasonably Achievable. This principle is even more valid for the good quality images obtained with modern ultrasound scanners. Reducing the transmitted power instead of reducing signal gain is a good example of using this principle in daily scanning.

---

[24]Salvesen KÅ, Eik-Nes SH. Ultrasound during pregnancy and birthweight, childhood malignancies and neurological development. *Ultrasound Med Biol* **1999** 25:1025–1031.

[25]Salvesen KÅ. Ultrasound and left-handedness: a sinister association? *Ultrasound Obstet Gynecol* **2002** 19:217–221.

[26]Salvesen KÅ, Vatten LJ, Eik-Nes SH, Hugdahl K, Bakketeig LS. Routine ultrasonography in utero and subsequent handedness and neurological development. *Br Med J* **1993** 307:159–164.

[27]Torloni MR, Vedmedovska N, Merialdi M, Betran AP, Allen T, Gonzalez R, Platt LD. Safety of ultrasonography in pregnancy: WHO systematic review of the literature and meta-analysis. *Ultrasound Obstet Gynecol* **2009** 33:599–608.

[28]Houston LE, Odibo AO, Macones GA. The safety of obstetrical ultrasound: a review. *Prenat Diagn* **2009** 29:1204–1212.

[29]Phillips RA, Stratmeyer ME, Harris GR. Safety and U.S. Regulatory considerations in the nonclinical use of medical ultrasound devices. *Ultrasound Med Biol* **2010** 36:1224–1228.

# 8

# Bubble physics

The density and compressibility parameters of blood cells hardly differ from those of plasma. Therefore, blood cells are poor scatterers in the clinical diagnostic frequency range. Since imaging blood flow and measuring organ perfusion are desirable for diagnostic purposes, markers should be added to the blood to differentiate between blood and other tissue types. Such markers must be acoustically active in the medical ultrasonic frequency range.

Figure 8.1 shows the resonance frequencies of free and encapsulated gas microbubbles as a function of their equilibrium radius. The resonance frequencies of encapsulated microbubbles lie slightly higher than those of free gas bubbles, but clearly well within the clinical diagnostic range, too. Based on their acoustic properties, microbubbles are well suited as an ultrasound contrast agent.

In this chapter, microbubble behaviour in an ultrasound field is explored, with special attention to the influence of the bubble shell.

## 8.1 Hollow sphere

Consider a thin-shelled sphere in equilibrium. Assume $p_\mathrm{s}$ to be the difference between the internal pressure and the ambient pressure, generally referred to as the surface pressure. For any cross-sectional area $A$ through the centre of the sphere, the following force balance must hold:

$$p_\mathrm{s}\, A = \sigma\, S, \tag{8.1}$$

where $S$ is the path around the area and $\sigma$ is the surface tension. Introducing the radius $R$ yields

$$p_\mathrm{s}\left(\pi R^2\right) = \sigma\left(2\pi R\right), \tag{8.2}$$

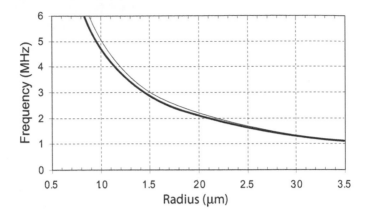

Figure 8.1: Resonance frequencies of free (bold line) and lipid-encapsulated (thin line) air microbubbles in water as a function of equilibrium radius.

which equates to

$$p_s = \frac{2\sigma}{R}. \tag{8.3}$$

Hence, the smaller the bubble, the higher the difference between the internal pressure and the ambient pressure. Since fluids are forced to flow from a location with a higher pressure to a location with a lower pressure, a bubble without an impenetrable solid shell cannot exist in true equilibrium.

## 8.2   Cavitation threshold

Now, consider a polytropic gas bubble in an infinite liquid. The following unstable equilibrium can be formulated:

$$p_g + p_v = p_0 + \frac{2\sigma}{R_0}, \tag{8.4}$$

where $p_g$ is the gas pressure, $p_v$ is the vapour pressure, $p_0$ is the ambient pressure, and $R_0$ is the quasi-equilibrium radius of the bubble.

If buoyancy and gas diffusion are slow compared with a change in ambient pressure,

$$p_g V^\gamma = \text{constant}, \tag{8.5}$$

where $V$ is the bubble volume and $\gamma$ is the ratio of specific heats of the gas. For air, $\gamma = 1.4$ is a good approximation. Substituting (8.4) for the gas pressure gives for any situation $n$

$$\left( p_0 - p_v + \frac{2\sigma}{R_0} \right) V_0^\gamma = p_n V_n^\gamma, \tag{8.6}$$

where $V_0$ is the quasi-equilibrium bubble volume. Changing the liquid pressure instantaneously, so that the liquid pressure at the bubble wall is $p_L$, gives

$$\left(p_0 - p_v + \frac{2\sigma}{R_0}\right) V_0^\gamma = \left(p_L - p_v + \frac{2\sigma}{R}\right) V^\gamma. \tag{8.7}$$

For a perfectly spherical bubble,

$$\left(p_0 - p_v + \frac{2\sigma}{R_0}\right) \left(\frac{4}{3}\pi R_0^3\right)^\gamma = \left(p_L - p_v + \frac{2\sigma}{R}\right) \left(\frac{4}{3}\pi R^3\right)^\gamma, \tag{8.8}$$

which can be rewritten as

$$p_L = \left(p_0 - p_v + \frac{2\sigma}{R_0}\right) \left(\frac{R_0}{R}\right)^{3\gamma} + p_v - \frac{2\sigma}{R}. \tag{8.9}$$

If the sonicating frequency is much lower than the bubble resonance frequency, the pressure in the liquid changes very slowly and uniformly compared with the natural time scale of the microbubble. The radius of a bubble $R$ in response to quasistatic changes in the liquid pressure is described by (8.9). Figure 8.2 shows the right-hand side of (8.9), for different $R_0$.

For each curve, there exists a minimum $(p_{cr}, R_{cr})$, where $R_{cr}$ is the critical radius and $p_{cr}$ is the critical quasi-isostatic pressure. The region to the right-hand side of the critical radius represents unstable equilibrium conditions. If the liquid pressure is lowered until it reaches a value below $p_{cr}$, no equilibrium radius exists, resulting in explosive growth of the bubble, much larger than $R_0$, hence the term cavitation threshold. The ambient pressure eventually increases again, during the ultrasonic compression phase, causing the bubble to collapse violently.

The critical radius is computed, knowing that, in $(R_{cr}, p_{cr})$,

$$\frac{\partial p_L}{\partial R} = 0. \tag{8.10}$$

Substituting the right-hand side of (8.9) for $p_L$ gives

$$-3\gamma \left(p_0 - p_v + \frac{2\sigma}{R_0}\right) \frac{R_0^{3\gamma}}{R_{cr}^{3\gamma+1}} + \frac{2\sigma}{R_{cr}^2} = 0, \tag{8.11}$$

which equates to

$$R_{cr} = \left[\frac{3\gamma}{2\sigma}\left(p_0 - p_v + \frac{2\sigma}{R_0}\right) R_0^{3\gamma}\right]^{\frac{1}{3\gamma-1}}, \tag{8.12}$$

from which the critical pressure follows:

$$p_{cr} = -p_0 + p_v - \frac{(6 - 2\gamma)\sigma}{3\gamma R_{cr}}, \tag{8.13}$$

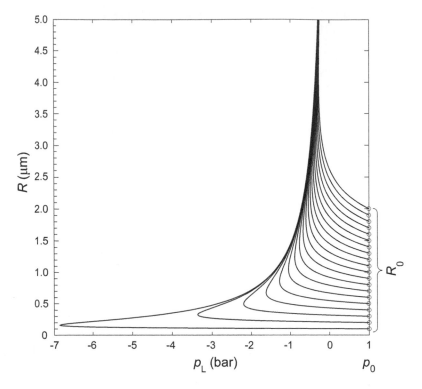

Figure 8.2: Solutions of (8.9) for different equilibrium radii $0.1 \leq R_0 \leq 2.0\,\mu\text{m}$, taking $p_0 = 1\,\text{atm}$, $\gamma = 1.4$, and $\sigma = 0.072\,\text{kg s}^{-2}$.

or, as a function of $R_0$,

$$p_{\text{cr}} = -p_0 + p_v - \frac{(6 - 2\gamma)\sigma}{3\gamma} \left[ \frac{2\sigma}{3\gamma \left( p_0 - p_v + \frac{2\sigma}{R_0} \right)} \right]^{3\gamma-1}. \tag{8.14}$$

If the situation is isothermal and if vapour pressure can be neglected, for bubbles of radius $R_0 \ll \frac{2\sigma}{p_0}$,

$$p_{\text{cr}} \approx -p_0 - 0.77 \frac{\sigma}{R_0}. \tag{8.15}$$

The critical radius, also referred to as the Blake radius, has been approximated by

$$R_{\text{cr}} \approx 2R_0. \tag{8.16}$$

During the initial part of the collapse the acceleration $\ddot{R}$ is negative. This sign changes as the gas inside the bubble begins to be compressed, and the rebound begins.

# 8.3 Fundamental equation of bubble dynamics

Consider an empty cavity with initial radius $R_0$ that expands or contracts to $R$, owing to a difference between the pressure in the liquid at the bubble wall and the pressure in the liquid at infinity $p_L - p_0^\infty$. Here, we take $p_0^\infty = p_0$. In time $\Delta t$, the liquid mass flowing across a surface outside the bubble with radius $r$ must equal the mass displaced by the expanding or contracting bubbles surface, i.e.,

$$4\pi r^2 \, \rho \dot{r} \, \Delta t = 4\pi R^2 \, \rho \dot{R} \, \Delta t. \tag{8.17}$$

Hence, the particle velocity in the liquid can be expressed in terms of $r$, $R$, and $\dot{R}$:

$$\dot{r} = \frac{R^2 \dot{R}}{r^2}. \tag{8.18}$$

The work done by an expanding or contracting bubble must equal the kinetic energy of the surrounding liquid:

$$\int_{R_0}^{R} (p_L - p_0) \, 4\pi R^2 \, \mathrm{d}R = \frac{1}{2} \int_{R}^{\infty} \dot{r}^2 \rho \, 4\pi r^2 \, \mathrm{d}r. \tag{8.19}$$

Substituting (8.18) for $\dot{r}$ simplifies the kinetic energy of the liquid to

$$E_k = 2\rho \int_{R}^{\infty} \frac{R^4 \dot{R}^2}{r^2} \, \mathrm{d}r = 2\pi \, \rho \, R^3 \dot{R}^2. \tag{8.20}$$

Now the following equality should be noted:

$$\frac{\partial}{\partial R} \left( \dot{R}^2 \right) = \frac{1}{\dot{R}} \frac{\partial \dot{R}^2}{\partial t} = 2\ddot{R}, \tag{8.21}$$

so that (8.19) can be differentiated to $R$. This results in the fundamental equation of bubble dynamics:

$$\frac{p_L - p_0}{\rho} = R\ddot{R} + \frac{3}{2}\dot{R}^2. \tag{8.22}$$

If a bubble is subjected to a driving function $P(t)$, (8.22) changes to

$$\frac{p_L - p_0 - P(t)}{\rho} = R\ddot{R} + \frac{3}{2}\dot{R}^2. \tag{8.23}$$

For a polytropic gas bubble, (8.9) is substituted for $p_L$:

$$R\ddot{R} + \frac{3}{2}\dot{R}^2 = \frac{1}{\rho}\left[ \left(p_0 - p_v + \frac{2\sigma}{R_0}\right)\left(\frac{R_0}{R}\right)^{3\gamma} + p_v - \frac{2\sigma}{R} - p_0 - P(t) \right]. \tag{8.24}$$

## 8.4   Pressure radiated by a bubble

To compute the acoustic pressure radiated by a bubble at any point in the liquid, consider the equation of motion (4.7):

$$\frac{1}{\rho}\frac{\partial p}{\partial r} = -\frac{\partial \dot{r}}{\partial t} - \dot{r}\frac{\partial \dot{r}}{\partial r}. \tag{8.25}$$

Integrating over $r$ gives

$$\int_r^\infty \frac{1}{\rho}\frac{\partial p}{\partial r}\,\mathrm{d}r = -\int_r^\infty \frac{\partial \dot{r}}{\partial t}\,\mathrm{d}r - \int_r^\infty \dot{r}\frac{\partial \dot{r}}{\partial r}\,\mathrm{d}r, \tag{8.26}$$

which can be solved by substituting (8.18) for $\dot{r}$:

$$\frac{p(r,t) - p_0}{\rho} = -\frac{\partial}{\partial t}\left(\frac{R^2 \dot{R}}{r}\right) - \frac{1}{2}\frac{R^4 \dot{R}^2}{r^4}. \tag{8.27}$$

This is actually a representation of Bernoulli's theorem,

$$\frac{p(r,t) - p_0^\infty}{\rho} = -\frac{\partial \Phi}{\partial t} - \frac{1}{2}v^2, \tag{8.28}$$

where $v$ is the particle velocity and $\Phi$ is the velocity potential

$$\Phi = -\int_r^\infty \dot{r}\,\mathrm{d}r. \tag{8.29}$$

The equation of motion in the liquid (8.27) can be further simplified to

$$\frac{p(r,t) - p_0}{\rho} = -\frac{2R\dot{R}^2 + R^2\ddot{R}}{r} - \frac{1}{2}\frac{R^4\dot{R}^2}{r^4}. \tag{8.30}$$

In the far field, at distances $r \gg R$,

$$\frac{p(r,t) - p_0}{\rho} = -\frac{2R\dot{R}^2 + R^2\ddot{R}}{r}. \tag{8.31}$$

## 8.5   Viscous fluids

The viscosity $\eta$ of a Newtonian viscous fluid is by definition the ratio of stress and rate of strain $\dot{\varepsilon}$. In viscous fluids, the relations (2.61) and (2.74) do not apply. It should be noted that the principal stresses have been defined as positive for expanding media, as opposed to the definitions in fluid physics and acoustics. If we take a hydrostatic stress $p$, for an incompressible liquid,

$$p_\mathrm{L} = -p - 2\eta\dot{\varepsilon}_\mathrm{r}, \tag{8.32}$$

where $\dot{\varepsilon}_r$ is the radial rate of strain. Using (8.18), the radial rate of strain can be expressed in terms of $r$ and $R$:

$$\dot{\varepsilon}_r = \frac{\partial \dot{r}}{\partial r} = \frac{\partial}{\partial r}\left(\frac{R^2 \dot{R}}{r^2}\right) = -\frac{2R^2 \dot{R}}{r^3}, \tag{8.33}$$

which at the bubble surface $(r = R)$ becomes

$$\dot{\varepsilon}_r = -\frac{2\dot{R}}{R}, \tag{8.34}$$

Combining (8.22), (8.32), and (8.34) results in

$$\frac{1}{\rho}\left(p_L - p_0 - \frac{4\eta \dot{R}}{R}\right) = R\ddot{R} + \frac{3}{2}\dot{R}^2. \tag{8.35}$$

Introducing a driving function $P(t)$ gives an equation similar to (8.24) for a polytropic gas bubble:

$$R\ddot{R} + \frac{3}{2}\dot{R}^2 = \frac{1}{\rho}\left[\left(p_0 - p_v + \frac{2\sigma}{R_0}\right)\left(\frac{R_0}{R}\right)^{3\gamma} + p_v - \frac{2\sigma}{R} - \frac{4\eta \dot{R}}{R} - p_0 - P(t)\right]. \tag{8.36}$$

This is the Rayleigh–Plesset equation. Note that the Rayleigh–Plesset equation can only be applied if the liquid is incompressible and if the gas is polytropic.

Figure 8.3 shows radius–time curves of two microbubbles subjected to continuous sine pressure waves with low, moderate, and high amplitudes. Both bubbles oscillate linearly at MI = 0.01. With increasing driving amplitude, asymmetries in radial excursion and expansion time rise, especially for the bigger bubble, which is closer to the resonance size. At MI = 0.8, both bubbles expand to a factor of the initial size, followed by a rapid collapse for the smaller bubble. The bigger bubble demonstrates collapses at MI = 0.18 and higher.

## 8.6  Oscillations

The Rayleigh–Plesset equation describes highly nonlinear radially symmetric bubble oscillations, but at low acoustic driving amplitudes, the behaviour is linear. At such low amplitudes, a bubble behaves like a mass–spring–dashpot system and (8.36) is just another way of writing (3.38), where the replacive mass

$$m = 4\pi R_0^3 \rho, \tag{8.37}$$

the linear angular resonance frequency

$$\omega_0 = \left(\frac{1}{R_0\sqrt{\rho}}\right)\sqrt{3\gamma\left(p_0 - p_v + \frac{2\sigma}{R_0}\right) + p_v - \frac{2\sigma}{R_0} - \frac{4\eta^2}{\rho R_0^2}}, \tag{8.38}$$

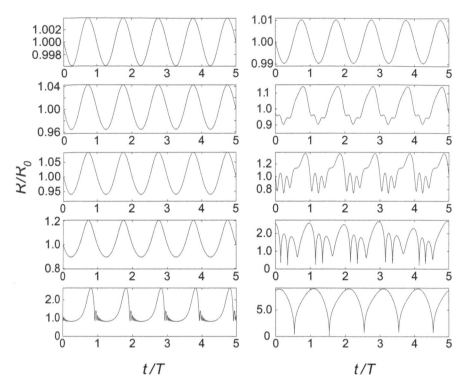

Figure 8.3: Simulated radius–time curves (radius $R$ normalised by equilibrium radius $R_0$, time $t$ normalised by period $T$) of ultrasound contrast microbubbles with $0.55\,\mu$m (*left column*) and $2.3\,\mu$m (*right column*) equilibrium radii. The modelled ultrasound field was a continuous sine wave with a frequency of $0.5\,$MHz and acoustic amplitudes corresponding to (top–bottom) MI = 0.01, 0.10, 0.18, 0.35, and 0.80. Reprinted with permission from Postema M, Gilja OH. Ultrasound-directed drug delivery. *Curr Pharm Biotechnol* **2007** 8:355–361.

and the (viscous) damping

$$2\zeta = \frac{16\pi\eta R_0}{m\omega_0} = \frac{4\eta}{\rho\omega_0 R_0^2}. \tag{8.39}$$

The damping of a bubble pulsation is determined by the acoustic radiation, the heat conduction, and the liquid viscosity. For microbubbles under sonication at typical medical frequencies $> 1\,$MHz, viscous damping is dominant, as is evident from (4.147). For an encapsulated microbubble, the presence of a shell has to be taken into account, by adding an extra damping parameter $\zeta_{\mathrm{s}}$. From (3.63) we know that the excursion of a forced damped harmonic oscillator has a phase angle difference $\phi$ with the driving field. Figure 8.4 shows three curves of the phase angle differences $(\phi + \pi)$ between a damped radially oscillating bubble and an incident 2-MHz sound field, as a function of $R_0$. The curves have been computed for a free microbubble, a SonoVue$^{TM}$ contrast microbubble, and an Albunex$^{®}$ contrast microbubble. With increasing shell stiffness, the bubble resonance size increases. At resonance, the bubble oscillates $\frac{3}{2}\pi$ rad out of phase with the sound field. For bubble greater than resonance, the phase angle difference approaches $2\pi$ rad, so that the bubble oscillates in phase with the sound field. Below resonance size, the phase difference is still greater than $\pi$, and approaches $\frac{3}{2}\pi$ for $R_0$ much smaller than resonance size. Since the damping due to the liquid viscosity $\zeta_v \propto R^{-2}$, the phase difference approaches $\frac{3}{2}\pi$ for a free bubble radius $R_0 \ll 1\,\mu$m. The approach to $\frac{3}{2}\pi$ below the minimum value of the phase difference is stronger with the contrast bubbles, because $\zeta_s \propto R^{-3}$. As the damping becomes greater, the phase transition around resonance becomes less abrupt, as Figure 8.4 demonstrates.[1]

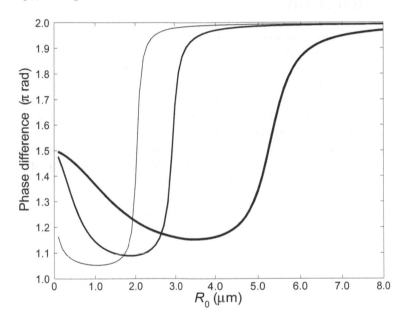

Figure 8.4: Phase angle difference $(\phi + \pi)$ between a damped radially oscillating bubble and an incident 2-MHz sound field, as a function of equilibrium radius $R_0$. The thin line represents a free bubble, the medium line a SonoVue$^{TM}$ microbubble, and the thickest line an Albunex$^{®}$ microbubble.

The spherically symmetric oscillating behaviour of ultrasound contrast agent microbubbles has been described with models based on the Rayleigh–Plesset equation, modified for the presence of an encapsulating shell. Generally, the presence of blood has a relatively small effect on bubble dynamics. To give an

[1]Postema M, Schmitz G. Ultrasonic bubbles in medicine: influence of the shell. *Ultrason Sonochem* **2007** 14:438–444.

indication of the vast amount of existing models: Qin *et al.* defined 16 separate dynamic bubble model classes.[2] The reason for the high number of existing models is the fact that most physical properties of encapsulated microbubbles cannot actually be measured, so that pseudo-material properties have to be chosen when predicting ultrasound contrast agent microbubble behaviour. Examples of such pseudo-material properties are shell elasticity parameters and shell friction parameters.

If the ultrasonic driving pressure is sufficiently high, the nonlinear microbubble response results in harmonic dispersion, which not only produces harmonics with frequencies that are integer multiples of $\omega$ (superharmonics) but also subharmonics with frequencies less than $\omega$ of the form $m\omega/n$, where $\{m, n\} \in \mathbb{N}$.

## 8.7 Disruption

At low acoustic amplitudes (mechanical index $\text{MI} < 0.1$), microbubbles pulsate linearly. At high amplitudes ($\text{MI} > 0.6$), their elongated expansion phase is followed by a violent collapse. During the collapse phase, when the kinetic energy of the bubble surpasses its surface energy, a bubble may fragment into a number of smaller bubbles. Fragmentation has been exclusively observed with contrast agents with thin, elastic shells. Fragmentation is the dominant disruption mechanism for these bubbles.

During the initial part of the collapse, the acceleration $\ddot{R}$ is negative. This sign changes as the gas inside the bubble begins to be compressed, and the rebound begins. Provided that surface instabilities have grown big enough to allow for break-up, microbubble fragmentation has been expected and observed close to this moment, when $\ddot{R} = 0$. This has been confirmed by means of high-speed photography. The occurrence of fragmentation has been associated with inertial cavitation.

The number of fragments, $N$, into which a microbubble breaks up, is related to the dominant spherical harmonic oscillation mode $n$ by[3]

$$N \approx n^3. \tag{8.40}$$

Mode 2 oscillations have been observed with lipid-encapsulated microbubbles, leading to fragmentation into 8 newly formed microbubbles.

Let us consider a single spherically symmetric microbubble with an inner radius $R_i$ and an outer radius $R$, a shell density $\rho_s$, negligible translation, in an infinite fluid with density $\rho$. The kinetic energy of such a microbubble can be approximated by

$$E_k \approx 2\pi \rho R^3 \dot{R}^2 + 2\pi \rho_s R_i^3 \dot{R}_i^2 \left(1 - \frac{R_i}{R}\right). \tag{8.41}$$

[2]Qin S, Caskey CF, Ferrara KW. Ultrasound contrast agent microbubbles in imaging and therapy: physical principles and engineering. *Phys Med Biol* **2009** 54:R27–R57.

[3]Brennen CE. Fission of collapsing cavitation bubbles. *J Fluid Mech* **2002** 472:153–166.

Knowing that, for microbubbles with monolayer lipid shells, $\frac{R_i}{R} < 0.01$ and $\rho_s = 1.15 \times 10^3$ kg m$^{-3}$, and for blood, $\rho = 1.05 \times 10^3$ kg m$^{-3}$, (8.41) can be reduced to (8.20).

The surface free energy $E_s$ of a single encapsulated bubble is given by

$$E_s = 4\pi R_i^2 \sigma_1 + 4\pi R^2 \sigma_2 , \qquad (8.42)$$

where $\sigma_1$ and $\sigma_2$ denote the surface tension coefficient for the inner and outer interface, respectively. For our microbubbles with monolayer lipid shells, we consider a single interface model, using the effective surface tension $\sigma$:

$$\sigma = \sigma_1 + \sigma_2 . \qquad (8.43)$$

After fragmentation, the resulting microbubble fragments contain more surface free energy $\sum_i E_{f,i}$ than the single bubble prior to fragmentation:

$$\sum_{i=1}^{N} E_{f,i} \approx \tfrac{4}{3}\pi R_{f,m}^2 \sigma N \approx \tfrac{4}{3}\pi R^2 \sigma N^{\frac{1}{3}} = N^{\frac{1}{3}} E_s , \qquad (8.44)$$

where $R_{f,m}$ is the mean fragment radius. Neglecting the elastic energy of the shell and the internal energy of the gas core, it can be assumed that fragmentation will only occur if:

$$E_k > \sum_{i=1}^{N} E_{f,i} - E_s. \qquad (8.45)$$

Although asymmetric shape bubble oscillations have been observed, within the size range of ultrasound contrast agent bubbles, spherical harmonic modes higher than 2 can be neglected.

For microbubbles of radius $R_0$ with a thick, stiff shell, such as Quantison$^{\text{TM}}$, $\max(R(t)) \ll R_0$. Thick-shelled bubbles have demonstrated gas release during a high-amplitude ultrasonic cycle. The increased pressure difference between the inside and outside of the bubble during the expansion phase of the wave causes the shell to be stretched across the critical deformation, resulting in its mechanical cracking. The released bubble has an oscillation amplitude much higher than an encapsulated bubble of the same size.

Figure 8.5 shows the ultrasound-induced release of gas from an albumin-encapsulated microbubble, driving the bubble at 0.5 MHz with a peak-negative acoustic pressure of 0.8 MPa.[4] The frames cover one full ultrasonic cycle (2 $\mu$s). This acoustic pressure is well within the clinical diagnostic range. Gas is seen to escape from the thick-shelled microbubble with a 4.3 $\mu$m diameter in the third frame, in the beginning of the rarefaction phase of the ultrasound. The shell itself is too rigid to expand. The released gas expands to a diameter of 12.3 $\mu$m in the eighth frame, after which it contracts. In the eleventh frame, the free gas

---

[4]Postema M, Bouakaz A, Versluis M, de Jong N. Ultrasound-induced gas release from contrast agent microbubbles. *IEEE Trans Ultrason Ferroelectr Freq Control* **2005** 52:1035–1041.

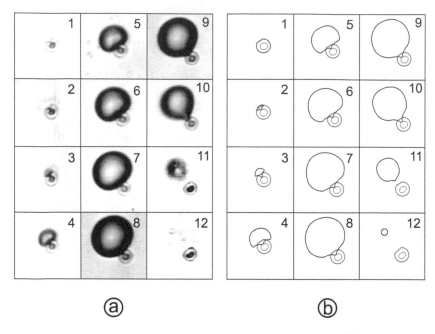

Figure 8.5: Gas release from the upper left of a Quantsion™ microbubble during a single ultrasonic cycle (a), and a schematic representation thereof (b). During the rarefaction phase (starting frame 2), gas escapes until it reaches a maximum (frame 8). During the subsequent contraction, the free gas bubble is seen detached from the shell (frames 11 and 12). Each frame corresponds to a $19 \times 19 \, (\mu m)^2$ area. Inter-frame times are $0.1 \, \mu s$. Reprinted with permission from Postema M, van Wamel A, ten Cate FJ, de Jong N. High-speed photography during ultrasound illustrates potential therapeutic applications of microbubbles. *Med Phys* **2005** 32:3707–3711.

microbubble, which has been subjected to motion blur, appears to be detached from the encapsulated microbubble. In the twelfth frame, the gas is hardly visible, owing to the compression phase of the ultrasound.

On the contrary, microbubbles with a thin, highly elastic monolayer lipid shell, like SonoVue™, have been observed to expand to more than ten-fold their initial surface areas during rarefaction. The shell behaves like an elastic membrane that ruptures under relatively small strain. By the time of maximal expansion, therefore, the shell has ruptured, leaving newly formed clean free interfaces.

## 8.8 Diffusion

In a steady fluid, gas diffusion is given by Fick's law:

$$\frac{\partial C}{\partial t} = D \left( \frac{\partial^2 C}{\partial r^2} + \frac{2}{r} \frac{\partial C}{\partial r} \right), \tag{8.46}$$

where $C$ is the mass concentration of the dissolved gas and $D$ is the dissolution constant. We introduce

$$u(r,t) = r \left( C - C_\mathrm{s} \right) \tag{8.47}$$

and the boundary condition

$$u(r,0) = r \left( C_\mathrm{i} - C_\mathrm{s} \right), \tag{8.48}$$

where $C_\mathrm{i}$ is the initial mass concentration of the dissolved gas and $C_\mathrm{s}$ is the saturation concentration in the liquid at the bubble wall. Then,

$$\frac{\partial u}{\partial t} = D \frac{\partial^2 u}{\partial r^2}. \tag{8.49}$$

The solution of this ordinary differential equation is

$$u(r,t) = u(r,0) \operatorname{erf}(z), \tag{8.50}$$

where

$$z = \frac{r}{2\sqrt{Dt}}. \tag{8.51}$$

The error function $\operatorname{erf}(z)$ is defined by

$$\operatorname{erf}(z) = \frac{2}{\sqrt{\pi}} \int_0^z e^{-\xi^2} \mathrm{d}\xi \tag{8.52}$$

and can be written as an asymptotic series

$$\operatorname{erf}(z) = 1 - \frac{e^{-z^2}}{\sqrt{\pi}} \sum_{n=0}^{\infty} \frac{(-1)^n (2n-1)!!}{2^n} z^{-(2n+1)} = 1 - \frac{e^{-z^2}}{\sqrt{\pi}} \left( z^{-1} - \frac{z^{-3}}{3} + \cdots \right). \tag{8.53}$$

Substituting for $u(r,0)$, (8.50) now becomes

$$u(r,t) = \frac{2r \left( C_\mathrm{i} - C_\mathrm{s} \right)}{\sqrt{\pi}} \int_0^{\frac{r}{2\sqrt{Dt}}} e^{-\xi^2} \mathrm{d}\xi. \tag{8.54}$$

Using the asymptotic series for $\operatorname{erf}(z)$ and the Taylor series for $e^z$, it follows that, at $r = R$,

$$\left( \frac{\partial u}{\partial r} \right)_R = \left( C_\mathrm{i} - C_\mathrm{s} \right) \left( 1 + \frac{R}{\sqrt{\pi Dt}} \right) \tag{8.55}$$

and, consequently,

$$\left(\frac{\partial C}{\partial r}\right)_R = (C_i - C_s)\left(\frac{1}{R} + \frac{1}{\sqrt{\pi Dt}}\right). \tag{8.56}$$

At the bubble wall, the mass flow through the surface equals the diffusion:

$$D\left(\frac{\partial C}{\partial r}\right)_R = \frac{1}{4\pi R^2}\frac{dm}{dt} = \frac{1}{4\pi R^2}\frac{d}{dt}\left(\frac{4}{3}\pi R^3 \rho_g\right) \tag{8.57}$$

or

$$4\pi R^2 \dot{R}\rho_g = 4\pi R^2 D\left(\frac{\partial C}{\partial r}\right)_R \tag{8.58}$$

where $\rho_g$ is the density of the gas. Substituting (8.56) yields the bubble wall velocity during dissolution:

$$\dot{R} = \frac{D(C_i - C_s)}{\rho_g}\left(\frac{1}{R} + \frac{1}{\sqrt{\pi Dt}}\right). \tag{8.59}$$

In this equation, $\rho_g$ is a function of $R$. Combining (4.22) and (8.4) rephrases the ideal gas law for a gas bubble:

$$p_0 + \frac{2\sigma}{R} = \frac{\rho_g \mathcal{R}T}{M}, \tag{8.60}$$

so that $\rho_g$ is expressed in terms of known parameters:

$$\rho_g(R) = \frac{M}{\mathcal{R}T}(p_0 - p_v) + \frac{2M\sigma}{\mathcal{R}T}\frac{1}{R} = \rho_g(\infty) + \frac{2M\sigma}{\mathcal{R}T}\frac{1}{R}, \tag{8.61}$$

where $\rho_g(\infty)$ is the density of the gas under the same conditions of pressure and temperature with a gas–liquid interface of zero curvature.[5] Substituting (8.61) into (8.57) and computing the mass diffusion rephrases (8.59) as

$$\dot{R} = \frac{D(C_i - C_s)}{\rho_g(\infty) + \frac{4}{3}\frac{M\sigma}{\mathcal{R}T}\frac{1}{R}}\left(\frac{1}{R} + \frac{1}{\sqrt{\pi Dt}}\right) \tag{8.62}$$

or

$$\dot{R} = \frac{D\mathcal{R}T(C_i - C_s)}{M}\frac{1}{p_0 - p_v + \frac{4}{3}\frac{\sigma}{R}}\left(\frac{1}{R} + \frac{1}{\sqrt{\pi Dt}}\right). \tag{8.63}$$

The concentration of gas at the bubble wall $C_s$ is related to the internal gas pressure by

$$C_s = k_g^{-1}p_g = k_g^{-1}\left(p_0 - p_v + \frac{2\sigma}{R}\right), \tag{8.64}$$

where $k_g$ is Henry's constant defined in terms of the mass concentration of the gas. The saturation concentration of the gas is, by definition,[6]

$$C_0 = k_g^{-1}p_0. \tag{8.65}$$

[5] Epstein PS, Plesset MS. On the stability of gas bubbles in liquid–gas solutions. *J Chem Phys* **1950** 18:1505–1509.
[6] Eller A, Flynn AG. Rectified diffusion during nonlinear pulsations of cavitation bubbles. *J Acoust Soc Am* **1965** 37:493–503.

Hence, the concentration of gas at the bubble wall is related to the saturation concentration in the liquid by

$$C_s = C_0 \left( 1 - \frac{p_v}{p_0} + \frac{2\sigma}{p_0 R} \right). \tag{8.66}$$

Equation (8.63) now reduces to

$$\dot{R} = \frac{D\mathcal{R}TC_0}{Mp_0} \left( \frac{\frac{C_i}{C_0} - 1 + \frac{p_v}{p_0} - \frac{2\sigma}{Rp_0}}{1 - \frac{p_v}{p_0} + \frac{4}{3}\frac{\sigma}{Rp_0}} \right) \left( \frac{1}{R} + \frac{1}{\sqrt{\pi Dt}} \right), \tag{8.67}$$

which can be simplified to

$$\dot{R} = DL \left( \frac{\frac{C_i}{C_0} - 1 + \frac{p_v}{p_0} - \frac{2\sigma}{Rp_0}}{1 - \frac{p_v}{p_0} + \frac{4}{3}\frac{\sigma}{Rp_0}} \right) \left( \frac{1}{R} + \frac{1}{\sqrt{\pi Dt}} \right), \tag{8.68}$$

where $L$ is Ostwald's solubility coefficient.[7] If a hydrostatic overpressure $\Delta p$ is introduced, the dissolution can be readily derived in a similar fashion:

$$\dot{R} = DL \left( \frac{\frac{C_i}{C_0} - 1 + \frac{p_v}{p_0} - \frac{\Delta p}{p_0} - \frac{2\sigma}{Rp_0}}{1 - \frac{p_v}{p_0} + \frac{\Delta p}{p_0} + \frac{4}{3}\frac{\sigma}{Rp_0}} \right) \left( \frac{1}{R} + \frac{1}{\sqrt{\pi Dt}} \right). \tag{8.69}$$

Figure 8.6 shows diameter–time curves of free dissolving nitric oxide gas microbubbles at two different ambient pressures. The dissolution process of a 2-$\mu$m microbubble takes less than 2.5 ms. Increasing the ambient pressure slightly decreases the dissolution times.

## 8.9 Radiation forces

### 8.9.1 Travelling sound wave

Consider a pressure gradient $\nabla p$ across a bubble of volume $V$. The force acting on the bubble must be

$$F = -V\nabla p. \tag{8.70}$$

In an acoustic field, the pressure gradient constantly changes. Hence, we consider the average force acting on the bubble, following the analysis by Leighton:[8]

$$\langle F \rangle = - \langle V\nabla p \rangle. \tag{8.71}$$

Now, consider a plane single-frequency (monotonous) progressive wave in the $x$-direction, for which the pressure deviation from the ambient constant value is described by (4.35):

$$p = P_A \cos(\omega t - kx) \tag{8.72}$$

---

[7]Bouakaz A, Frinking PJA, de Jong N, Bom N. Noninvasive measurement of the hydrostatic pressure in a fluid-filled cavity based on the disappearance time of micrometer-sized free gas bubbles. *Ultrasound Med Biol* **1999** 25:1407–1415.

[8]Leighton TG. *The Acoustic Bubbles*. London: Academic Press **1994**.

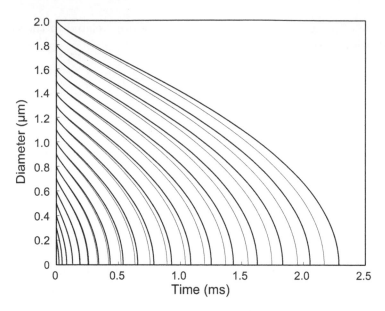

Figure 8.6: Diameter–time curves of dissolving nitric oxide gas bubbles at atmospheric pressure (bold lines) and 100 mmHg overpressure (thin lines), respectively. Reprinted with permission from Postema M, Bouakaz A, ten Cate FJ, Schmitz G, de Jong N, van Wamel A. Nitric oxide delivery by ultrasonic cracking: some limitations. *Ultrasonics* **2006** 44:e109–e113.

and

$$\nabla p = -kP_A \sin(\omega t - kx), \tag{8.73}$$

where $P_A$ is the acoustic pressure amplitude, $k$ is the wave number, and $\omega$ is the angular driving frequency. At small acoustic amplitudes, a bubble oscillates linearly:

$$R(t) = R_0 - \xi \cos(\omega t - kx - \phi), \tag{8.74}$$

where $\xi$ is the bubble oscillation amplitude and $\phi$ is the phase difference between the sound field and the bubble. The volumetric change is then approximated by

$$\begin{aligned}
V(t) &= \tfrac{4}{3}\pi \left[ R_0 - \xi \cos(\omega t - kx - \phi) \right]^3 \\
&= \tfrac{4}{3}\pi \left[ R_0^3 - 3R_0^2 \xi \cos(\omega t - kx - \phi) + 3R_0\xi^2 \cos^2(\omega t - kx - \phi) \right. \\
&\qquad \left. - \xi^3 \cos^3(\omega t - kx - \phi) \right] \\
&\approx V_0 \left[ 1 - \tfrac{3\xi}{R_0} \cos(\omega t - kx - \phi) \right].
\end{aligned} \tag{8.75}$$

Hence, the average force acting on the bubble is

$$\langle F \rangle = -\left\langle V_0 k P_A \left[ 1 - \frac{3\xi}{R_0} \cos(\omega t - kx - \phi) \right] \sin(\omega t - kx) \right\rangle. \tag{8.76}$$

Making use of $\sin A \cos(A + B) = \frac{1}{2} \sin 2A \cos B - \sin^2 A \sin B$, this becomes

$$\langle F \rangle = -V_0 k P_{\mathrm{A}} \left( \langle \sin(\omega t - kx) \rangle + \tfrac{3\xi}{R_0} \langle \sin^2(\omega t - kx) \sin \phi \rangle \right.$$
$$\left. + \langle \sin(\omega t - kx) \cos(\omega t - kx) \cos \phi \rangle \right). \tag{8.77}$$

The uneven terms are averaged out, whereas $\langle \sin^2 A \rangle = \frac{1}{2}$, so that

$$\langle F \rangle = \frac{3 V_0 k P_{\mathrm{A}}}{2} \frac{\xi}{R_0} \sin \phi. \tag{8.78}$$

Substituting (3.63) for $\phi$ and taking into account that $\sin \arctan x = \frac{x}{\sqrt{1+x^2}}$ gives:

$$\langle F \rangle = \frac{3 V_0 k P_{\mathrm{A}}}{2} \frac{\xi}{R_0} \frac{2\zeta \frac{\omega}{\omega_0}}{\sqrt{\left(1 - \left(\frac{\omega}{\omega_0}\right)^2\right)^2 + \left(2\zeta \frac{\omega}{\omega_0}\right)^2}}. \tag{8.79}$$

This force, acting in the direction of the sound field, is called the primary radiation force.

## 8.9.2 Standing sound wave

Consider a bubble in a standing sound wave

$$p = 2 P_{\mathrm{A}} \cos \omega t \cos kx \tag{8.80}$$

and

$$\nabla p = -k P_{\mathrm{A}} \sin(\omega t - kx). \tag{8.81}$$

At small acoustic amplitudes, the radius is then given by

$$R(t) = R_0 - \xi \cos kx \cos(\omega t - \phi). \tag{8.82}$$

Analogous to (8.75), the volumetric change is approximated by

$$V(t) \approx V_0 \left[ 1 - \frac{3\xi}{R_0} \cos kx \cos(\omega t - \phi) \right]. \tag{8.83}$$

Consequently, the average force acting on the bubble is

$$\langle F \rangle = -\left\langle 2 V_0 k P_{\mathrm{A}} \left[ 1 - \frac{3\xi}{R_0} \cos kx \cos(\omega t - \phi) \right] \sin kx \cos \omega t \right\rangle. \tag{8.84}$$

Again, the uneven terms are averaged out, so that

$$\langle F \rangle = \frac{3 V_0 k P_{\mathrm{A}}}{2} \frac{\xi}{R_0} \sin 2kx \cos \phi. \tag{8.85}$$

Substituting (3.63) for $\phi$ and taking into account that $\cos \arctan x = \frac{1}{\sqrt{1+x^2}}$ gives

$$\langle F \rangle = \frac{3V_0 k P_A \sin 2kx}{2} \frac{\xi}{R_0} \frac{1 - \left(\frac{\omega}{\omega_0}\right)^2}{\sqrt{\left(1 - \left(\frac{\omega}{\omega_0}\right)^2\right)^2 + \left(2\zeta \frac{\omega}{\omega_0}\right)^2}}. \tag{8.86}$$

This force, acting in the direction of the nodes and anti-nodes of the sound field, is called the primary Bjerknes force.

### 8.9.3  Radiation forces between bubbles

Consider an object in a sound field that causes a fluid acceleration $\dot{v}$ at the position of a bubble of interest. Defining $\dot{u}$ as the acceleration of the bubble, the net acceleration of the bubble relative to the fluid is $\dot{u} - \dot{v}$. This relative acceleration causes a drag force on the bubble $-\frac{1}{2}\rho V (\dot{u} - \dot{v})$, where $\frac{1}{2}\rho V$ is the apparent mass of a moving bubble. Following Leighton's derivation, the net force on the bubble is

$$F = \rho V \dot{v} - \frac{1}{2}\rho V (\dot{u} - \dot{v}) = \rho_g(t) V \dot{u}, \tag{8.87}$$

from which an expression for $\dot{u}$ immediately follows:

$$\dot{u} = \frac{3V\dot{v}}{V + 2V\frac{\rho_g}{\rho}}. \tag{8.88}$$

If the mass of the gas inside the bubble is constant,

$$\rho_g V = \rho_{0,g} V_0, \tag{8.89}$$

where $\rho_{0,g}$ is the density of the gas bubble in quasi-equilibrium. We assume the bubble oscillates linearly according to

$$V(t) = V_0 - \Delta V \cos \omega t, \tag{8.90}$$

where $\Delta V = 4\pi R^2 \xi$. We substitute this for $V$ and the density ratio f for $\frac{\rho_{0,g}}{\rho}$ in (8.88):

$$\frac{\dot{u}}{\dot{v}} = \frac{3(V_0 - \Delta V \cos \omega t)}{(1 + 2f)V_0 - \Delta V \cos \omega t}. \tag{8.91}$$

Using $\frac{1}{1-x} = 1 + x + x^2 + x^3 + ...$, this can be simplified to

$$\frac{\dot{u}}{\dot{v}} = \frac{3}{1 + 2f}\left(1 - \frac{\Delta V \cos \omega t}{V_0}\right)\left(1 + \frac{\Delta V \cos \omega t}{(1 + 2f)V_0}\right)$$

$$\approx \frac{3}{1 + 2f}\left(1 - \frac{2f}{1 + 2f}\frac{\Delta V}{V_0} \cos \omega t\right). \tag{8.92}$$

Now, consider that the object causing the fluid acceleration $\dot{v}$ is a bubble "1" at distance $r$ from the bubble of interest "2". If $V_1$ is the volume of bubble 1 at quasi-equilibrium, $V_2$ is the volume of bubble 2 at quasi-equilibrium, $\Delta V_1$ is the volumetric expansion amplitude of bubble 1, and $\Delta V_2$ is the volumetric expansion amplitude of bubble 2, then, assuming small oscillation amplitudes, the instantaneous volume of bubble 1 is $V_1 - \cos(\omega t + \phi)$ and the instantaneous volume of bubble of 2 is $V_2 - \cos \omega t$, where $\phi$ is the difference in oscillation phase. We define $\rho_1$ and $\rho_2$ as the gas density at equilibrium of bubble 1 and 2, respectively. Similar to (8.87), the average force experienced by bubble 2 is

$$\langle F \rangle = \langle \rho V \dot{u} \rangle = \rho_2 V_2 \langle \dot{u} \rangle = \frac{3}{1+2f} \left\langle \dot{v} \rho_2 V_2 - \frac{6f}{(1+f)^2} \dot{v} \rho_2 \Delta V_2 \cos \omega t \right\rangle. \quad (8.93)$$

Considering that $V_1 = \frac{4}{3}\pi R_1^3$ and that $\dot{V}_1 = 4\pi R_1^2 \dot{R}_1$, (8.18) can be rewritten in terms of $V_1$:

$$v = \frac{R_1^2 \dot{R}_1}{r^2} = \frac{\dot{V}_1}{4\pi r^2} = \frac{\omega \Delta V_1 \sin(\omega t + \phi)}{4\pi r^2}, \quad (8.94)$$

so that

$$\dot{v} = \frac{\omega^2 \Delta V_1 \cos(\omega t + \phi)}{4\pi r^2}. \quad (8.95)$$

Inserting this in (8.93) results in

$$\langle F \rangle = \frac{3}{1+2f} \frac{\rho_2 \omega^2 \Delta V_1 V_2}{4\pi r^2} \langle \cos(\omega t + \phi) \rangle$$

$$- \frac{6f}{(1+f)^2} \frac{\rho_2 \omega^2 \Delta V_1 \Delta V_2}{4\pi r^2} \langle \cos \omega t \cos(\omega t + \phi) \rangle \quad (8.96)$$

$$= -\frac{3f}{(1+2f)^2} \frac{\rho_2 \omega^2 \Delta V_1 \Delta V_2}{4\pi r^2} \cos \phi.$$

This force is called the secondary radiation or secondary Bjerknes force. From (8.96), it immediately follows that bubbles that oscillate in phase ($\phi = 0$) attract each other and that bubbles that oscillate out of phase ($\phi = \pi$) repel each other.

## 8.10   Coalescence

To understand microbubble coalescence, one needs to comprehend the drainage of the liquid separating the bubble surfaces. Reynolds noted that the viscosity of a liquid can be determined by pressing two flat plates together, squeezing the liquid out, and measuring the drainage velocity.[9] Thus, he formulated an equation for the drainage velocity of a fluid between rigid surfaces. General theories on the coalescence of colliding bubbles and droplets have been based on liquid

---

[9]Reynolds O. On the theory of lubrication and its application to Mr. Beauchamp Tower's experiments, including an experimental determination of the viscosity of olive oil. *Philos Trans Roy Soc A* **1886** 177:157–234.

film drainage.[10,11] Droplet coalescence finds applications in fuel ignition research and aerosol studies, whereas the research on bubble coalescence focuses on thin film physics and foam stability. This Section explores ultrasound-induced coalescence of microbubbles. Controlled coalescence has potential applications in the clinical field.[12]

Theories on bubble coalescence are generally based on the collision of unencapsulated bubbles or droplets, approaching each other at constant velocity. During expansion, microbubbles may also come into contact with each other, resulting in coalescence or bounce. We discriminate the following stages in the coalescence mechanism, optically observed in Figure 8.7 and schematically represented in Figure 8.8. Initially, two bubbles approach collision while expanding (Figure 8.8(a)). Prior to contact, there may be a flattening of the adjacent bubble surfaces, trapping liquid in between (Figure 8.7(a), Figure 8.8(b)). This trapped liquid drains (Figure 8.7(b), Figure 8.8(c)) until the separation reaches a critical thickness. An instability mechanism results in rupture of the separation (Figure 8.8(d)) and the formation of a merged bubble (Figure 8.7(c)). After coalescence the resulting bubble will have an ellipsoidal shape (Figure 8.7(d), Figure 8.8(e)). Owing to surface tension, it will relax to a spherical shape. When the contact time is less than the time needed for film drainage, the bubbles bounce off each other.[13] We define bubble coalescence as the fusing of two or more bubbles into a single bubble. The process begins with the flattening of the bubble surfaces and is considered finished when the resulting bubble has a spherical shape.

### 8.10.1  Flattening of the interface

Flattening of the opposing bubble surfaces occurs because the liquid inertia overcomes the capillary pressure, as described in earlier work on colliding bubbles with constant volumes. For colliding bubbles, flattening happens if the bubble system has a Weber number We $\gtrsim$ 0.5.[14] The Weber number for two colliding bubbles with radii $R_1$ and $R_2$, respectively, is given by the inertial force relative to the surface tension force:

$$\mathrm{We} = \frac{\rho u^2}{\frac{\sigma}{R_\mathrm{m}}} \, , \tag{8.97}$$

[10]Kralchevsky PA, Danov KD, Ivanov IB. Thin liquid film physics. In: Prud'homme R, Khan S, eds., *Foams, Theory, Measurements and Applications*. New York: Marcel Dekker **1996** 1–98.

[11]Narsimhan G, Ruckenstein E. Structure, drainage, and coalescence of foams and concentrated emulsions. In: Prud'homme R, Khan S, eds., *Foams, Theory, Measurements and Applications*. New York: Marcel Dekker **1996** 99–187.

[12]Postema M, Marmottant, Lancée CT, Hilgenfeldt S, de Jong N. Ultrasound-induced microbubble coalescence. *Ultrasound Med Biol* **2004** 30:1337–1344.

[13]Chaudhari RV, Hofmann H. Coalescence of gas bubbles in liquids. *Rev Chem Eng* **1994** 10:131–190.

[14]Duineveld PC. Bouncing and coalescence phenomena of two bubbles in water. In: Blake JR, Boulton-Stone JM, Thomas NH, eds., *Bubble Dynamics and Interface Phenomena*. Volume 23 of *Fluid mechanics and its applications*. Dordrecht: Kluwer Academic Publishers **1994** 447–456.

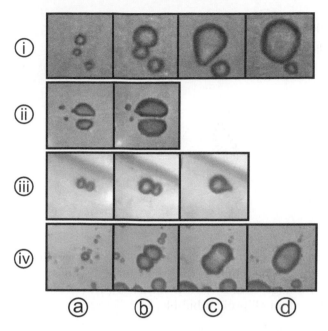

Figure 8.7: Optical images of stages of ultrasound-induced microbubble coalescence: (a) flattening of contact surfaces, (b) liquid film drainage, (c) forming of a merged bubble, (d) turning into an ellipsoidal bubble. Each frame in event (i) corresponds to a $21 \times 21 \, (\mu m)^2$ area. Each frame in events (ii)–(iv) corresponds to a $30 \times 30 \, (\mu m)^2$ area. Inter-frame times are $0.33 \, \mu s$. Reprinted with permission from Postema M, Marmottant P, Lancée CT, Hilgenfeldt S, de Jong N. Ultrasound-induced microbubble coalescence. *Ultrasound Med Biol* **2004** 30:1337–1344.

where $u$ is the relative approach velocity of the bubble walls, $\rho$ is the fluid density, $\sigma$ is the surface tension, and $R_\mathrm{m}$ is the mean bubble radius, which is defined by

$$\frac{2}{R_\mathrm{m}} = \frac{1}{R_1} + \frac{1}{R_2}. \tag{8.98}$$

Consider the Weber number criterium for approaching walls of expanding bubbles. Then, for bubbles with a constant centre-to-centre distance, $u = \dot{R}_1 + \dot{R}_2$. If the Weber number is low, bubble coalescence will always occur, without flattening of the adjacent surfaces prior to contact. In the high Weber number regime, coalescence is determined by a second step, after flattening: film drainage.

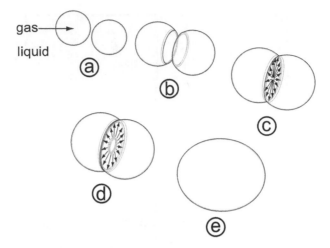

Figure 8.8: Schematic representation of stages of expanding bubble co-alescence: (a) bubble collision, (b) flattening of contact surfaces, (c) liquid film drainage until a critical thickness (d), (e) film rupture, and (f) formation of an ellipsoidal bubble.

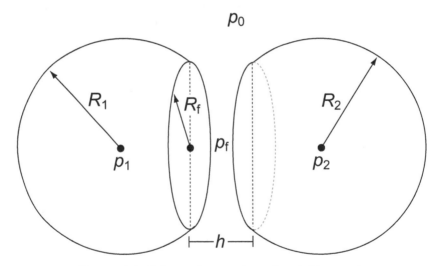

Figure 8.9: Schematic overview of variables used.

## 8.10.2 Film drainage

Consider two bubbles with radii $R_1$ and $R_2$, and internal pressures $p_1$ and $p_2$, respectively, assumed spherical everywhere with the exception of a flattened interface that separates them through a liquid film of thickness $h$ (*cf.* Figure 8.9). The drainage rate of the liquid film depends on the difference $(p + \Pi)$ between the film pressure $p_f$ and the liquid ambient pressure $p_0$. Here, $p$ is the difference in hydrodynamic pressure and $\Pi$ is the disjoining pressure in the film. We estimate the pressure in the film by the mean of pressures $p_1$ and $p_2$, since the parallel film surfaces lead to equal pressure differences towards both bubbles:

$$p + \Pi = p_f - p_0 = \frac{1}{2}(p_1 + p_2) - p_0 = \sigma\left(\frac{1}{R_1} + \frac{1}{R_2}\right) \equiv p_{LY}, \qquad (8.99)$$

where $p_{LY}$ is the Laplace–Young film pressure. The disjoining pressure begins to slow down film thinning when $h$ drops below $0.1\,\mu$m, and becomes the dominant pressure term (usually owing to Van der Waals forces) when $h$ thins to about $10\,$nm.[15] The eventual coalescence of ultrasound contrast agent microbubbles is very fast compared with the film drainage time scales considered later. Therefore, we may neglect $\Pi$ and take $p$ equal to the Laplace–Young pressure for the films observed. As such, the pressure gradient determining the drainage velocity is independent of the ambient pressure.

We choose an $r$–$z$ coordinate system such that the film is symmetric around the plane $z = 0$ and the line $r = 0$, and that its boundaries are located at $z = \pm\frac{1}{2}h$ and $r = R_f$. The Laplace–Young pressure gradient drives liquid out of the film. The radial velocity of the liquid is described by a combination of a plug flow (present without any resistance to flow) and a laminar flow profile (in $z$) of Poiseuille type induced by resistance at the film interfaces.[16] The drainage of the liquid film can be parameterised by functions of these two contributions. Below, the two limiting cases of bubbles with no-slip interfaces and bubbles with free interfaces are analysed.

### 8.10.3 No-slip interfaces

In the presence of surfactant at sufficient surface concentration, the interfaces can be considered immobile (no-slip). In the case of no-slip interfaces, the interfacial tangential velocity is zero, so the plug flow contribution is zero, as shown in frame (a) of Figure 8.10.

The film drainage velocity for rigid radial surfaces (disks) is given by the Reynolds equation:[17]

$$-\frac{\partial h}{\partial t} = \frac{2\,p\,h^3}{3\,\eta\,R_f^2}. \qquad (8.100)$$

The drainage time, $\tau_d$, between the initial film thickness $h_i$ and the critical film thickness $h_c$ can be determined by integration of (8.100):

$$\int_{h_i}^{h_c} -\frac{dh}{h^3} = \int_0^{\tau_d} \frac{2\,p}{3\,\eta\,R_f^2}\,dt. \qquad (8.101)$$

[15]Marrucci G. A theory of coalescence. *Chem Eng Sci* **1969** 24:975–985.

[16]Klaseboer E, Chevaillier JP, Gourdon C, Masbernat O. Film drainage between colliding drops at constant approach velocity: experiments and modeling. *J Colloid Interf Sci* **2000** 229:274–285.

[17]Sheludko A. Thin liquid films. *Advan Colloid Interf Sci* **1967** 1:391–464.

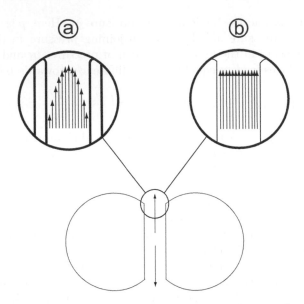

Figure 8.10: Schematic flow profiles between no-slip interfaces (a) and of free interfaces (b).

Flattening takes place when

$$\dot{R}_1 + \dot{R}_2 \gg \frac{\partial h}{\partial t},$$ (8.102)

whereas the flat film drainage happens in the next stage, when

$$\dot{R}_1 \approx \dot{R}_2 \approx 0.$$ (8.103)

Thus, during drainage, we may take $p$ and $R_f$ constant over time. Then we obtain

$$\tau_d = \frac{3\eta R_f^2}{4p h_c^2}\left(1 - \frac{h_c^2}{h_i^2}\right).$$ (8.104)

If $h_c^2 \ll h_i^2$ the drainage time can be approximated by

$$\tau_d \approx \frac{3\eta R_f^2}{4p h_c^2}.$$ (8.105)

### 8.10.4   Free interfaces

In the case of free interfaces, the Poiseuille contribution to the drainage flow becomes negligible, and the drainage is inertial, as shown in frame (b) of Figure 8.10. The film drainage velocity for free radial surfaces is given by the equation[18]

$$-\frac{\partial h}{\partial t} = \sqrt{\frac{8\,p}{\rho}}\,\frac{h}{R_{\mathrm f}}\,.$$ (8.106)

Note that the viscous term is absent. Similarly to the no-slip case, making the same quasi-static assumptions with regards to $p$ and $R_{\mathrm f}$, the drainage time can be approximated by

$$\tau_{\mathrm d} \approx R_{\mathrm f}\,\sqrt{\frac{\rho}{8\,p}}\,\log\left(\frac{h_{\mathrm i}}{h_{\mathrm c}}\right)\,.$$ (8.107)

These drainage times are much smaller than those for the no-slip situation, and depend only logarithmically on both the initial and the critical film thickness.

### 8.10.5  Film rupture

The disjoining pressure induces rupture by amplifying surface perturbations. These are initialised by either thermal fluctuations or by capillary waves.[19] For thermal perturbations of a gas bubble in the micrometre range, the initial perturbation will be on the order of $\sqrt{\frac{kT}{\sigma}}$, where k is Boltzmann's constant and $T$ is the absolute temperature, in our situation approximately 300 K. Hence, the initial thermal perturbation is lower than 1 nm.

A film gradually thins to a critical thickness at which it either ruptures due to a local instability or at which it attains an equilibrium thickness. These critical thicknesses dependent of surfactant concentration and film radius. They lie in the range 20 nm $< h_{\mathrm c} <$ 40 nm for film radii 60 $\mu$m $< R_{\mathrm f} <$ 160 $\mu$m.[20]

For ultrasound contrast agent film radii ($R_{\mathrm f} <$ 10 $\mu$m), we may assume critical thicknesses around 10 nm, knowing that below 10 nm Van der Waals forces become very strong and rapid rupture of the film (and thus coalescence) ensues. Because of the weak dependence on film thickness, predictions from (8.107) for coalescence time scales can be quite accurate even without precise knowledge of $h_{\mathrm i}$ and $h_{\mathrm c}$.

## 8.11  Jetting

The jetting phenomenon for cavitation bubbles can be described as follows.[21] Consider an oscillation bubble. Let's define an infinite boundary to the right-hand side of the bubble. During sonication, at the moment of maximal expansion (*cf.* Figure 8.11b1), the pressure inside the bubble is much lower than

[18]Kirkpatrick RD, Lockett MJ. The influence of approach velocity on bubble coalescence. *Chem Eng Sci* **1974** 29:2363–2373.

[19]Sharma A, Ruckenstein E. Critical thickness and lifetimes of foams and emulsions: role of surface wave-induced thinning. *J Colloid Interf Sci* **1987** 119:14–29.

[20]Angarska JK, Dimitrova BS, Danov KD, Kralchevsky PA, Ananthapadmanabhan KP, Lips A. Detection of the hydrophobic surface force in foam films by measurements of the critical thickness of film rupture. *Langmuir* **2004** 20:1799–1806.

[21]Postema M, van Wamel A, ten Cate FJ, de Jong N. High-speed photography during ultrasound illustrates potential therapeutic applications of microbubbles. *Med Phys* **2005** 32:3707–3711.

the ambient pressure, causing the bubble to collapse. The radial water flow is retarded by the boundary. Therefore, the pressure at the right bubble wall is less than the pressure at the right wall during the whole collapse phase and the bubble becomes elongated perpendicular to the boundary. The pressure gradient leads to different accelerations of the left and right bubble walls and therefore to a movement of the centre of the bubble towards the boundary during collapse. As the bubble collapses, the fluid volume to the left of the bubble is accelerated and focussed, leading to the formation of a liquid jet directed towards the boundary. This jet hits the right-hand-side bubble wall, causing a funnel-shaped protrusion (*cf.* Figure 8.11b2) and finally impacts the boundary.

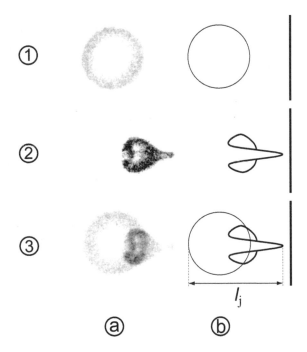

Figure 8.11: Two high-speed photographic frames (a1,2) and an overlaid image thereof (a3) of microjetting — a microbubble acting as a microsyringe — and a schematic representation of this phenomenon (b). On the verge of microjetting (1, thin line), the microbubble has a diameter of $17\,\mu$m. During microjetting (2), liquid protrudes through the right side of the microbubble, over a length of $l_j = 26\,\mu$m. The jet is represented by the bold curve. The time between the two frames is $0.33\,\mu$s. Reprinted with permission from Postema M, van Wamel A, ten Cate FJ, de Jong N. High-speed photography during ultrasound illustrates potential therapeutic applications of microbubbles. *Med Phys* **2005** 32:3707–3711.

Empirical relations exist between bubble radius, jet length, and pressure at the tip of jets. The radius of the jet $R_j$ is related to the radius of the bubble on the verge of collapsing $R_c$ by[22]

$$\frac{R_j}{R_c} \approx 0.1. \tag{8.108}$$

The length of the jet, $l_j$, defined as the full travel path of the protruded liquid, is related to $R_c$ by[23]

$$\frac{l_j}{R_c} \approx 3. \tag{8.109}$$

From these two ratios, the amount of liquid within the jet, $V_j$, can be estimated:[24]

$$V_j \approx 0.1\, R_c^3. \tag{8.110}$$

The impact of a jet on a surface generates a high pressure region. The pressure in this region has been referred to as water-hammer.[25] For a perfectly plastic impact, the water-hammer pressure of a cavitation jet is approximately[26]

$$p_{wh} \approx \frac{1}{2}\,\rho\,c\,v_j, \tag{8.111}$$

where $p_{wh}$ is the water-hammer pressure and $v_j$ is the jet velocity.

When administering microbubbles in the bloodstream, vessel walls are the boundaries to which ultrasound-induced jets are to be targeted. From high-speed optical observations of microjetting through ultrasound contrast agent microbubbles, it has been computed that the pressure at the tip of the jet is high enough to penetrate any human cell.[27] Therefore, it has been speculated whether liquid jets might act as microsyringes, delivering a drug to a region of interest.

Of influence on the occurrence of all the above-mentioned phenomena are (a) the ultrasonic parameters: transmit frequency, acoustic amplitude, pulse length, pulse repetition rate and transmit phase; (b) the ultrasound contrast agent composition: the composition of the shell, the bubble sizes, the size distribution and the gas; (c) the physical properties of the medium: viscosity, surface tension, saturation. Table 8.1 gives an overview of the nonlinear phenomena that have been observed with ultrasound contrast agents, the type of ultrasound contrast agent in which they have occurred, and the minimum acoustic regime required.

---

[22]Kodama T, Takayama K. Dynamic behavior of bubbles during extracorporeal shock-wave lithotripsy. *Ultrasound Med Biol* **1998** 24:723–738.

[23]Ohl CD, Ikink R. Shock-wave-induced jetting of micron-size bubbles. *Phys Rev Lett* **2003** 90:214502.

[24]Ohl CD, Ory E. Aspherical bubble collapse — comparison with simulations. In Lauterborn W, Kurz T, eds., *Nonlinear Acoustics at the Turn of the Millennium*. New York: American Institute of Physics **2000** 393–396.

[25]Cook SS. Erosion by water-hammer. *Proc Roy Soc London A* **1928** 119:481–488.

[26]de Haller P. Untersuchungen über die durch Kavitation hergerufenen Korrosionen. *Schweiz Bauzeit* **1933** 101:243–246.

[27]Postema M, van Wamel A, Lancée CT, de Jong N. Ultrasound-induced encapsulated microbubble phenomena. *Ultrasound Med Biol* **2004** 30:827–840.

| Phenomenon | Schematic representation | Shell class[1] | Regime[2] |
|---|---|---|---|
| Translation | | I, II, III, IV | L, M, H |
| Fragmentation | | I, II | L, M, H |
| Coalescence | | I, II | L, M, H |
| Jetting | | I, II | H |
| Clustering | | II, III | L, M, H |
| Cracking | | II, III, IV | L, M, H |

[a]Microbubble shell classes: (I) free or released gas; (II) thin shells $< 10\,\text{nm}$; (III) thick shells $< 500\,\text{nm}$; (IV) very thick shells $> 500\,\text{nm}$.

[b]Acoustic regimes: low (L) for $MI < 0.3$; medium (M) for $0.3 < MI < 0.7$; high (H) for $MI > 0.7$.

Table 8.1: Nonlinear phenomena and their occurrence regimes.

# 9

# CEUS and sonoporation

## with Odd Helge Gilja and Annemieke van Wamel

In this chapter, contrast-enhanced ultrasound (CEUS) and ultrasound contrast agent microbubble adjustments for drug delivery including sonoporation are described. Although hundreds of papers have been published on this subject, we will just briefly touch upon the topic using a physics and engineering approach.

## 9.1  Commercial ultrasound contrast agents

Microbubbles that are used for ultrasonic imaging purposes are termed ultrasound contrast agents. After intravascular injection of an agent into the circulation, microbubbles pass the site of interest that is being examined by the clinician. Upon sonication, the microbubbles generate a microbubble-characteristic response, which is detected by the ultrasound scanner used. The resulting sonographic image can then be interpreted by the clinician.

The development of ultrasound contrast agents has gone through several generations.[1] Table 9.1 gives an overview of the ultrasound contrast agents that are most commonly used in imaging research.[2] Free microbubbles represent generation 0. These bubbles rapidly dissolve owing to diffusion.

To prevent rapid dissolution, generation 0 microbubbles were large: they could have diameters up to $80\,\mu$m. These large sizes would prevent the microbubbles from passing the lung capillaries; and thus also contributed to rapid

---

[1]Krestan C. Ultraschallkontrastmittel: Substanzklassen, Pharmakokinetik, klinische Anwendungen, Sicherheitsaspekte. *Radiologe* **2005** 45:513–519.

[2]Postema M, Schmitz G. Bubble dynamics involved in ultrasonic imaging. *Expert Rev Mol Diagn* **2006** 6:493–502.

circulation clearance. Extended circulation time has been established with first generation ultrasound contrast agents. These consist of air bubbles encapsulated by a stabilising shell. With mean diameters below 6 $\mu$m, these bubbles are small enough to pass through capillaries.

If an ultrasound contrast agent contains perfluorocarbon gas rather than air, the microbubbles will first swell, due to the diffusion of dissolved gases into the bubbles, and then dissolve. The low diffusion rate of high-molecular-weight perfluorocarbons prolongs microbubble presence from seconds to minutes.[3] Often, the surface of the bubble shell has a negative charge, to prolong its presence in target tissue.[4]

Ultrasound contrast agents can be designed to specifically target a biomarker molecule, often a glycoprotein or receptor molecule, thus facilitating ultrasonic molecular imaging. Third generation ultrasound contrast agents consist of microbubbles with such special targeting (functionalised) shell properties. Owing to primary radiation forces, microbubbles can be forced to translate away from the transducer, toward the vessel walls, increasing the success rate of targeting.

[3]Schutt EG, Klein DH, Mattrey RM, Riess JG. Injectable microbubbles as contrast agents for diagnostic ultrasound imaging: the key role of perfluorochemicals. *Angew Chem Int Ed* **2003** 42:3218–3235.

[4]Fisher NG, Christiansen JP, Klibanov A, Taylor RP, Kaul S, Lindner JR. Influence of microbubble surface charge on capillary transit and myocardial contrast enhancement. *J Am Coll Cardiol* **2002** 40:811–819.

| Agent | Original developer | Shell | Gas/vapor | Mean diameter ($\mu m$) |
|---|---|---|---|---|
| *First generation*[1] | | | | |
| Albunex | Mallinckrodt Inc. | Albumin | Air | 4.3 |
| Infoson | Mallinckrodt Inc. | Albumin | Air | 4.3 |
| Levovist® | Schering AG | galactose/palmitic acid | Air | 2–3 |
| Sonovist® | Schering AG | Cyanoacrylate | $SF_6$ | 1–2 |
| *Second generation*[2] | | | | |
| BR14 | Bracco Diagnostics Inc. | Lipid | $C_4F_{10}$ | 2.5–3.0 |
| Definity® | Bristol-Myers Squibb Medical Imaging, Inc. | Lipid/surfactant | $C_3F_8$ | 1.1–3.3 |
| Imagent® | IMCOR Pharmaceuticals, Inc. | Lipid/surfactant | $C_6F_{14}/N_2$ | 6.0 |
| Optison™ | Molecular Biosystems Inc. | Albumin | $C_3F_8$ | 2.0–4.5 |
| Quantison™ | Andaris Ltd. | Albumin | Air | 3.2 |
| SonoVue® | Bracco Diagnostics Inc. | Lipid | $SF_6$ | 2.5 |
| *Third generation*[3] | | | | |
| AI-700 | Acusphere, Inc. | Poly-L-lactide co glycolide | $C_4F_{10}$ | 2 |
| CARDIOsphere® | POINT Biomedical Corp. | Polylactide/albumin | $N_2$ | 4.0 |
| EchoGen® | Sonus Pharmaceuticals, Inc. | Surfactant | $C_5F_{12}$ | 2–5 |
| MicroMarker™ | VisualSonics Inc. | Lipid | $C_4F_{10}/N_2$ | 2.5 |
| Sonazoid | Amersham Health | Lipid/surfactant | $C_4F_{10}$ | 2.4–3.6 |
| Targestar™ | Targeson Inc. | Lipid | $C_4F_{10}$ | 2.5 |

Table 9.1: The most common ultrasound contrast agents.*

[1]First generation: encapsulated air bubbles.
[2]Second generation: encapsulated low solubility gas bubbles.
[3]Third generation: particulated gas bubbles with controlled acoustic properties.
*Based on Postema M, Schmitz G. Bubble dynamics involved in ultrasonic imaging. *Expert Rev Mol Diagn* **2006** 6:493–502.

# 9.2 CEUS

Originally, ultrasound contrast studies were performed for left ventricular function and myocardial perfusion.[5] Nowadays, ultrasound contrast agents have, among others, been used in the following diagnostic techniques: imaging the heart,[6] vasculature (including vasa vasorum), liver, spleen, kidneys,[7] brain,[8] measuring tissue perfusion, ejection fractions,[9] detecting focal lesions in the liver, angiogenesis assessment,[10] characterising tumours, and detecting sites of inflammation.[11]

Figure 9.1 shows how perfused areas become clearly visible when an ultrasound contrast agent is administered.

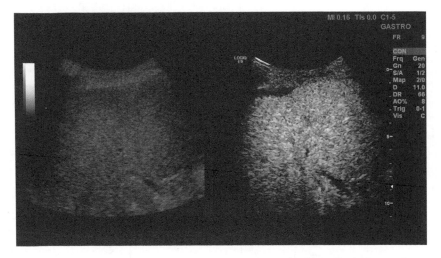

Figure 9.1: B-mode scans of the liver without (*left*) and with an ultrasound contrast agent present (*right*), using a dynamic 1–5-MHz probe.

Contrast-enhanced ultrasound (CEUS) represents a significant advancement in the evaluation of angiogenesis in cancers in the digestive system. Particularly,

[5]Becher H, Burns PN. *Handbook of Contrast Echocardiography: LV Function and Myocardial Perfusion*. Berlin: Springer **2000**.

[6]Miller AP, Nanda NC. Contrast echocardiography: new agents. *Ultrasound Med Biol* **2004** 30:425–434.

[7]Heynemann H, Jenderka KV, Zacharias M, Fornara P. Neue Techniken der Urosonographie. *Urologe* **2004** 43:1362–1370.

[8]Droste DW, Kaps M, Navabi DG, Ringelstein EB. Ultrasound contrast enhancing agents in neurosonology: principles, methods, future possibilities. *Acta Neurol Scand* **2000** 102:1–10.

[9]Mischi M, Jansen AHM, Kalker AACM, Korsten HHM. Identification of ultrasound contrast agent dilution systems for ejection fraction measurements. *IEEE Trans Ultrason Ferroelectr Freq Control* **2005** 52:410–420.

[10]Heppner P, Lindner JR. Contrast ultrasound assessment of angiogenesis by perfusion and molecular imaging. *Expert Rev Mol Diagn* **2005** 5:447–455.

[11]Lindner JR. Microbubbles in medical imaging: current applications and future directions. *Nature Rev Drug Discov* **2004** 3:527–532.

for the study of focal liver lesions, CEUS has been widely used for detection and characterisation of malignancy. The unique feature of CEUS of non-invasive assessment in real time of liver perfusion throughout the vascular phases has led to a great improvement in diagnostic accuracy of ultrasound, but also in guidance and evaluation of responses to therapy. Currently, CEUS is part of the state-of-the-art diagnostic work-up of focal liver lesions, resulting in safe and cost-effective patient management.

## 9.3 Some non-cardiac imaging applications

### 9.3.1 Liver

Ultrasonography is the most commonly used imaging modality worldwide for diseases of the liver. However, it has limited sensitivity in the detection of small tumour nodules. Moreover, ultrasonographic findings are often non-specific, as appearances of benign and malignant liver lesions overlap considerably. The introduction of microbubble contrast agents and the development of contrast-specific techniques have opened new prospects in liver ultrasonography. The advent of second-generation agents that enable continuous real-time contrast-enhanced imaging has been instrumental in improving the acceptance and reproducibility of the examination. With the publication of guidelines for the use of contrast agents in liver ultrasonography by the European Federation of Societies for Ultrasound in Medicine and Biology (EFSUMB),[12,13] contrast-enhanced ultrasonography is now routinely used in clinical practice.

As opposed to contrast media used with computed tomography (CT) and magnetic resonance (MR) imaging, ultrasound contrast agents can visualise the capillary net of the examined tissue, because CEUS is considerably more sensitive to very small amounts of contrast agent, even to single bubbles. Furthermore, because sonography is a dynamic method that is performed in real time, additional information about tissue perfusion can be deduced from the influx and washout of the contrast media, thus facilitating the differential diagnosis of tumours. In addition, signals from the microbubbles enable the visualisation of slow flow in microscopic vessels without Doppler-related artefacts. Various software packages have been developed to enable quantification of changes in contrast intensity and yield additional objective information over the entire course of the contrast examination.

---

[12] Albrecht T, Blomley M, Bolondi L, Claudon M, Correas JM, Cosgrove D, Greiner L, Jäger K, Jong ND, Leen E, Lencioni R, Lindsell D, Martegani A, Solbiati L, Thorelius L, Tranquart F, Weskott HP, Whittingham T. Guidelines for the use of contrast agents in ultrasound. January 2004. *Ultraschall Med* **2004** 25:249–256.

[13] Claudon M, Cosgrove D, Albrecht T, Bolondi L, Bosio M, Calliada F, Correas JM, Darge K, Dietrich C, D'Onofrio M, Evans DH, Filice C, Greiner L, Jäger K, Jong N, Leen E, Lencioni R, Lindsell D, Martegani A, Meairs S, Nolsøe C, Piscaglia F, Ricci P, Seidel G, Skjoldbye B, Solbiati L, Thorelius L, Tranquart F, Weskott HP, Whittingham T. Guidelines and good clinical practice recommendations for contrast enhanced ultrasound (CEUS) — update 2008. *Ultraschall Med* **2008** 29:28–44.

## 9.3.2 Pancreas

The pancreas, lying deep to the stomach and duodenum, is among the most inaccessible organs in the body for visualisation with ultrasonography. Hence, confirmation of pancreatic disease has remained a great challenge in clinical imaging. However, transabdominal ultrasonography has developed to be a useful tool in the differential diagnosis of pancreatic tumours because the technique is inexpensive, easy to perform, and widely available. Nevertheless, only after the introduction of second-generation contrast media has transabdominal sonography yielded results comparable to those of other diagnostic modalities. CEUS can be used to improve detection of pancreatic lesions or to characterise pancreatic lesions already visible with ultrasonography. Moreover, the staging of some pancreatic lesions can be improved by the use of contrast media. However, there is an important difference between a pancreatic CEUS study and the well-established liver CEUS study: the blood supply of the pancreas is entirely arterial and the enhancement of the gland begins almost together with the aortic enhancement. With CEUS the enhancement reaches its peak between 15 s and 20 s after injection of the ultrasound contrast agent. Accordingly, pancreatic tissue enhancement is earlier and shorter than that of the liver due to the absence of a venous blood supply like the portal vein for the liver. After a marked parenchymal enhancement in the early contrast-enhanced arterial phase, there is a progressive washout of contrast medium with gradual loss of echogenicity.[14–17]

## 9.3.3 Gastrointestinal tract

Colon cancer is one of the world's most commonly occurring malignancies. The main therapy is surgical resection. To diagnose colon cancer, endoscopy is the preferred method, but in many places around the world, X-ray is still applied. Using ultrasonography, the normal gastrointestinal (GI) wall is visualised as a layered structure consisting of five to nine layers, depending on transmitted frequency. When digestive cancers develop, the wall layers become blurred, wall thickness is increased, and the ultrasound appearance of the GI wall resembles a kidney, i.e., pseudo-kidney sign or target lesion. However, CEUS does not yet have a place in the work-up of patients with suspected colonic cancer.

On the contrary, in ultrasound imaging of inflammatory bowel diseases, CEUS appears to be a promising tool to visualise and quantify perfusion of the bowel wall layers. Recent studies indicate that CEUS measurement is as-

[14]Faccioli N, Crippa S, Bassi C, D'Onofrio M. Contrast-enhanced ultrasonography of the pancreas. *Pancreatology* **2009** 9:560–566.

[15]Dörffel Y, Wermke W. Neuroendocrine tumors: characterization with contrast-enhanced ultrasonography. *Ultraschall Med* **2008** 29:506–514.

[16]Faccioli N, D'Onofrio M, Malagò R, Zamboni G, Falconi M, Capelli P, Mucelli RP. Resectable pancreatic adenocarcinoma: depiction of tumoral margins at contrast-enhanced ultrasonography. *Pancreas* **2008** 37:265–268.

[17]Recaldini C, Carrafiello G, Bertolotti E, Angeretti MG, Fugazzola C. Contrast-enhanced ultrasonograpic findings in pancreatic tumors. *Int J Med Sci* **2008** 5:203–208.

sociated with other clinical parameters of disease activity in Crohn's disease. Accordingly, CEUS of bowel inflammation may be a significant future method for the diagnostic work-up of these patients and possibly also for monitoring effect of treatment.[18–21]

## 9.4 Molecular imaging

Dayton and Rychak define molecular imaging as the non-invasive application of an imaging modality to discern changes in physiology on a molecular level.[22] Although ultrasound contrast agents have been intended for perfusion imaging, they have proven useful in molecular imaging as well, after modification of the microbubble shell. Dayton and Rychak discern two targeting strategies: active targeting, in which a ligand specific for a particular molecular target is used, and passive targeting, in which the physicochemical properties of the agent are used to achieve retention at the target site. Molecular imaging and targeting have been reviewed elsewhere in depth. In summary, the main applications include the deep tissue detection of angiogenesis, inflammation, plaques, and thrombi, which are involved in most cardiovascular and malignant diseases.

## 9.5 Increased drug uptake

It has been proven by numerous groups that the cellular uptake (endocytosis) of drugs and genes is increased when the region of interest is under sonication, and even more so when an ultrasound contrast agent is present.[23] This increased uptake has been attributed to the formation of transient porosities in the cell membrane that are big enough for the transport of drugs into the cell. The transient permeabilisation and resealing of a cell membrane is called sonoporation.[24] The sonoporation-induced cellular uptake of markers with molecular

---

[18]Serra C, Menozzi G, Labate AM, Giangregorio F, Gionchetti P, Beltrami M, Robotti D, Fornari F, Cammarota T. Ultrasound assessment of vascularization of the thickened terminal ileum wall in Crohn's disease patients using a low-mechanical index real-time scanning technique with a second generation ultrasound contrast agent. *Eur J Radiol* **2007** 62:114–121.

[19]Migaleddu V, Scanu AM, Quaia E, Rocca PC, Dore MP, Scanu D, Azzali L, Virgilio G. Contrast-enhanced ultrasonographic evaluation of inflammatory activity in Crohn's disease. *Gastroenterology* **2009** 137:43–52.

[20]Quaia E, Migaleddu V, Baratella E, Pizzolato R, Rossi A, Grotto M, Cova MA. The diagnostic value of small bowel wall vascularity after sulfur hexafluoride-filled microbubble injection in patients with Crohn's disease. Correlation with the therapeutic effectiveness of specific anti-inflammatory treatment. *Eur J Radiol* **2009** 69:438–444.

[21]Nylund K, Hausken T, Gilja OH. Ultrasound and inflammatory bowel disease. *Ultrasound Q* **2010** 26:3–15.

[22]Dayton PA, Rychak JJ. Molecular ultrasound imaging using microbubble contrast agents. *Frontiers Biosci* **2007** 12:5124–5142.

[23]Postema M, Gilja OH. Ultrasound-directed drug delivery. *Curr Pharm Biotechnol* **2007** 8:355–361.

[24]Bao S, Thrall BD, Miller DL. Transfection of a reporter plasmid into cultured cells by sonoporation in vitro. *Ultrasound Med Biol* **1997** 23:953–959.

weights between 10 kDa and 3 MDa has been reported in several studies.[25,26] Schlicher *et al.* have shown that ultrasound-induced cavitation facilitates cellular uptake of macromolecules with diameters up to 56 nm.[27] Even solid spheres with a 100-nm diameter have been successfully delivered with the aid of sonoporation.[28] This implies that drug size is not a limiting factor for intracellular delivery. However, the pore opening times can be so short that, if the drug is to be effectively internalised, it should be released close to the cell membrane when poration occurs.[29]

## 9.6   Causes of sonoporation

There are five non-exclusive hypotheses for explaining the sonoporation phenomenon. These have been summarised in Figure 9.2.

It has been hypothesised that expanding microbubbles might push the cell membrane inward, and that collapsing bubbles might pull cell membranes outward.[30] These mechanisms require microbubbles to be present in the close vicinity of cells. A separate release mechanism should then ensure localised delivery. Although jetting only occurs in a high-MI regime, it is very effective in puncturing cell membranes.

Jetting has been observed through cells using ultrasound contrast agent microbubbles. However, the acoustic impedance of the solid cell substratum formed the boundary to which the jetting took place, not the cell itself.[31] Also, there has not been any proof yet of cell survival after jetting. In a separate study, the role of jetting was excluded as a dominant mechanism in sonoporation.[32]

If a microbubble is fixed to a membrane, the fluid streaming around the oscillating bubbles creates enough shear to rupture the membrane.[33] Here again, a separate release mechanism should then ensure localised delivery. Finally, it has been speculated that lipid-encapsulted microbubbles, in the compressed

[25] Miller DL, Bao S, Morris JE. Sonoporation of cultured cells in the rotating tube exposure system. *Ultrasound Med Biol* **1999** 25:143–149.

[26] Karshafian R, Samac S, Banerjee M, Bevan PD, Burns PN. Ultrasound-induced uptake of different size markers in mammalian cells. *Proc IEEE Ultrason Symp* **2005** 1:13–16.

[27] Schlicher RK, Radhakrishna H, Tolentino TP, Apkarian RP, Zarnitsyn V, Prausnitz MR. Mechanism of intracellular delivery by acoustic cavitation. *Ultrasound Med Biol* **2006** 32:915–924.

[28] Song J, Chappell JC, Qi M, VanGieson EJ, Kaul S, Price RJ. Influence of injection site, microvascular pressure and ultrasound variables on microbubble-mediated delivery of microspheres to muscle. *J Am Coll Cardiol* **2002** 39:726–731.

[29] Mehier-Humbert S, Bettinger T, Yan F, Guy RH. Plasma membrane poration induced by ultrasound exposure: implication for drug delivery. *J Control Release* **2005** 104:213–222.

[30] van Wamel A, Kooiman K, Harteveld M, Emmer M, ten Cate FJ, Versluis M, de Jong N. Vibrating microbubbles poking individual cells: drug transfer into cells via sonoporation. *J Control Release* **2006** 112:149–155.

[31] Prentice P, Cuschieri A, Dholakia K, Prausnitz M, Campbell P. Membrane disruption by optically controlled microbubble cavitation. *Nature Phys* **2005** 1:107–110.

[32] Postema M, Gilja OH. Jetting does not cause sonoporation. *Biomed Tech (Biomed Eng)* **2010** 55:S19–S20.

[33] Marmottant P, Hilgenfeldt S. Controlled vesicle deformation and lysis by single oscillating bubbles. *Nature* **2003** 423:153–156.

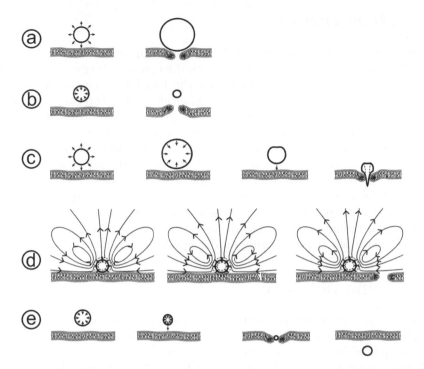

Figure 9.2: Possible mechanisms of sonoporation: (a) push, (b) pull, (c) jetting, (d) shear, (e) translation.

phase, translate through cell membranes or channels in the cell membrane. In case of therapeutic loading, the load would be delivered directly into the target cell.

The main advantage of the latter mechanism is that microbubble translation by means of ultrasonic radiation forces requires very low acoustic pressures. Hence, any potential damaging bioeffects due to inertial cavitation can be ruled out.

Although the fragmentation of therapeutic load-bearing microbubbles must release their loads, the actual actual drug or gene delivery is in this case a passive process, dependent on diffusion speed and release proximity to the target cells. Fragmenting microbubbles cannot create pores in cells, since fragmentation costs energy.

Without the presence of an agent, it has been assumed that sonoporation is caused by bubbles, which have been generated in the transducer focus as a result of inertial cavitation.[34,35]

[34]Miller DL, Nyborg WL. Theoretical investigation of the response of gas-filled micropores and cavitation nuclei to ultrasound. *J Acoust Soc Am* **1983** 73:1537–1544.

[35]Miller DL, Song J. Lithotripter shock waves with cavitation nucleation agents produce tumor growth reduction and gene transfer in vivo. *Ultrasound Med Biol* **2002** 28:1343–1348.

## 9.7 Drug carriers

Instead of just facilitating the transient opening up of cell membranes, a microbubble might also act as the vehicle itself to carry a drug or gene load to a perfused region of interest, in which case the load has to be released with the assistance of ultrasound. Apart from mixing ultrasound contrast agent with a therapeutic agent, several schemes have been proposed to combine microbubbles with a therapeutic load. The following seven microbubble structure classes for drug delivery have been discriminated:[36] (a) attachment to the outer shell surface; (b) intercalation between monolayer phospholipids; (c) incorporation in a layer of oil; (d) complexes with smaller particles (secondary carriers); (e) physical encapsulation in a polymer layer and coating with biocompatible material; (f) surface loading of protein-shelled microbubbles; (g) entire volume loading of protein-shelled microbubbles. The drugs are to be released at the site of interest during insonication,[37] presumably by disrupting the microbubble shell.

## 9.8 Gene delivery

The albumin shells of ultrasound contrast agents can bind proteins and oligonucleotides.[38] Local gene delivery of a virus vector attached to albumin-encapsulated microbubbles has been performed *in vivo*.[39]

It has been demonstrated *in vitro* that higher doses of deoxyribonucleic acid (DNA) were delivered during ultrasound insonication when the DNA was loaded on albumin-encapsulated microbubbles than when unloaded microbubbles were mixed with plasmid DNA. Amounts of DNA loading on microbubbles have been between $0.002$ (pg $\mu m^{-2}$)[40] and $2.4$ (pg $\mu m^{-2}$).[41]

## 9.9 Therapeutic gases

Instead of attached a drug to the encapsulation, therapeutic compounds in the gas phase could be encapsulated with thick shells, to keep them from dissolving. At the region of interest, the shell would be cracked with ultrasound, releasing

---

[36]Tinkov S, Bekeredjian R, Winter G, Coester C. Microbubbles as ultrasound triggered drug carriers. *J Pharm Sci* **2009** 98:1935–1961.

[37]Klibanov AL. Targeted delivery of gas-filled microspheres, contrast agents for ultrasound imaging. *Adv Drug Delivery Rev* **1999** 37:139–157.

[38]Porter TR, Xie F. Therapeutic ultrasound for gene delivery. *Echocardiography* **2001** 18:349–353.

[39]Shohet RV, Chen S, Zhou Y-T, Wang Z, Meidell RS, Unger RH, Grayburn PA. Echocardiographic destruction of albumin microbubbles directs gene delivery to the myocardium. *Circulation* **2000** 101:2554–2556.

[40]Christiansen JP, French BA, Klibanov AL, Kaul S, Lindner JR. Targeted tissue transfection with ultrasound destruction of plasmid-bearing cationic microbubbles. *Ultrasound Med Biol* **2003** 29:1759–1767.

[41]Frenkel PA, Chen S, Thai T, Shohet RV, Grayburn PA. DNA-loaded albumin microbubbles enhance ultrasound-mediated transfection in vitro. *Ultrasound Med Biol* **2002** 28:817–822.

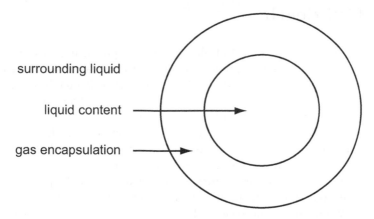

Figure 9.3: An antibubble consists of a liquid core encapsulated by a gas shell.

the gaseous content.[42,43] However, only few therapeutic compounds exist in the gaseous phase, *e.g.*, nitric oxide and several gaseous anaesthetics.

## 9.10 Antibubbles

A therapeutic agent inside the microbubble shell may react with the shell and dampen the bubble oscillations. Therefore, it might be more suitable to have the therapeutic agent in the core of the microbubble, separated from the shell by a gaseous layer. Incorporating a liquid drop containing drugs or genes inside an ultrasound contrast agent microbubble, however, is technically challenging. As opposed to bubbles, antibubbles consist of a liquid core surrounded by a gas encapsulation (*cf.* Figure 9.3). Such a droplet inside a bubble may be generated with the jetting phenomenon: the collapse of a bubble near a free surface produces a liquid jet, which may break up into one or several droplets.[44] Another option would be to stabilise the liquid core by means of a biodegradable skeleton attached to the microbubble shell.

[42]Bloch SH, Wan M, Dayton PA, Ferrara KW. Optical observation of lipid- and polymer-shelled ultrasound microbubble contrast agents. *Appl Phys Lett* **2004** 84:631–633.

[43]Dayton P, Morgan K, Allietta M, Klibanov A, Brandenburger G, Ferrara K. Simultaneous optical and acoustical observations of contrast agents. *Proc IEEE Ultrason Symp* **1997** 2:1583–1591.

[44]Postema M, ten Cate FJ, Schmitz G, de Jong N, van Wamel A. Generation of a droplet inside a microbubble with the aid of an ultrasound contrast agent: first result. *Lett Drug Des Discov* **2007** 4:74–77.

# 9.11   Cell death

It has been noted that if microbubbles can create pores, it is also possible to create severe cell and tissue damage. There is an inverse correlation between cell permeability and cell viability,[45,46] *i.e.*, not all cell membrane pores are temporary. This indicates that sonoporation is just a transitory membrane damage in the surviving cell. Cell lysis results from irreversible mechanical cell membrane damage,[47] which allows intracellular content to leak out. Recently, ultrasound-induced apoptosis has been observed with cancer cells *in vitro*,[48] and also in the presence of an ultrasound contrast agent.[49] Apart from situations where lysis is desired (sonolysis),[50] ultrasonic settings should be chosen such that cell lysis is minimal. Side effects observed are capillary rupture, haemorrhages, and dye extravasation.[51] These side effects, however, have been associated with relatively high microbubble concentrations, long ultrasonic pulse lengths, and high acoustic intensities.

# 9.12   High-intensity focussed ultrasound

High-intensity focussed ultrasound (HIFU) is a new non-invasive modality of cancer therapy using the thermal effects of ultrasound to ablate tumours.[52,53] The ultrasound energy is focussed in a small region inside the body, increasing the local temperature at which cell death occurs, whereas the temperature outside the focal region is low enough to prevent tissue damage. The small lesion size results in a long treatment time, usually several hours.[54] Experiments

---

[45]Miller DL, Dou C, Song J. DNA transfer and cell killing in epidermoid cells by diagnostic ultrasound activation of contrast agent gas bodies in vitro. *Ultrasound Med Biol* **2003** 29:601–607.

[46]Hallow DM, Mahajan AD, McCutchen TE, Prausnitz MR. Measurement and correlation of acoustic cavitation with cellular bioeffects. *Ultrasound Med Biol* **2006** 32:1111–1122.

[47]Feril Jr LB, Kondo T, Takaya K, Riesz P. Enhanced ultrasound-induced apoptosis and cell lysis by a hypotonic medium. *Int J Radiat Biol* **2004** 80:165–175.

[48]Watanabe A, Kawai K, Sato T, Nishimura H, Kawashima N, Takeuchi S. Apoptosis induction in cancer cells by ultrasound exposure. *Jap J Appl Phys* **2004** 43:3245–3248.

[49]Abdollahi A, Domhan S, Jenne JW, Hallaj M, Dell'Aqua G, Mueckenthaler M, Richter A, Martin H, Debus J, Ansorge W, Hynynen K, Huber PE. Apoptosis signals in lymphoblasts induced by focused ultrasound. *FASEB J* **2004** 18:1413–1414.

[50]Miller MW, Miller DL, Brayman AA. A review of in vitro bioeffects of inertial ultrasonic cavitation from a mechanistic perspective. *Ultrasound Med Biol* **1996** 22:1131–1154.

[51]Bekeredjian R, Grayburn PA, Shohet RV. Use of ultrasound contrast agents for gene or drug delivery in cardiovascular medicine. *J Am Coll Cardiol* **2005** 45:329–335.

[52]ter Haar GR. High intensity focused ultrasound for the treatment of tumors. *Echocardiography* **2001** 18:317–322.

[53]Liu Y, Kon T, Li C, Zhong P. High intensity focused ultrasound-induced gene activation in sublethally injured tumor cells in vitro. *J Acoust Soc Am* **2005** 118:3328–3336.

[54]Tung Y-S, Liu H-L, Wu C-C, Ju K-C, Chen W-S, Lin W-L. Contrast-agent-enhanced ultrasound thermal ablation. *Ultrasound Med Biol* **2006** 32:1103–1110.

have been performed *in vitro*[55] and *in vivo*,[56] demonstrating that the lesion size can be increased using an ultrasound contrast agent at the region of interest. Moreover, tissue damage can be produced more frequently, by lower acoustic intensities and shorter exposure, with an ultrasound contrast agent present.[57]

By making use of shock waves, even more destructive effects can be achieved. Techniques such as extracorporeal shock wave lithotripsy (ESWL) find applications in kidney stone fragmentation.[58,59]

# 9.13   Concluding remarks

It is a challenging task to quantify and predict which bubbly phenomenon occurs under which acoustic condition, and how these may be utilised in ultrasonic imaging. Aided by high-speed photography, our improved understanding of encapsulated microbubble behaviour will lead to more sophisticated detection and delivery techniques.

More sophisticated methods use quantitative approaches to measure the amount and the time course of bolus or reperfusion curves and have shown great promise in revealing effective tumour response to anti-angiogenic drugs in humans before tumour shrinkage occurs. These are beginning to be accepted into clinical practice. In the long term, targeted microbubbles for molecular imaging and eventually for directed anti-tumour therapy are expected to be tested.

In principle, in any perfused region that can be reached by ultrasound, ultrasound-directed drug delivery could be performed. However, since the ultrasonic fields used with diagnostic ultrasound scanners differ greatly per organ targeted, some regions will be far from ideal. The ultrasonic frequencies transmitted in endoscopy are much higher than the resonance frequencies of conventional ultrasound contrast agents. Therefore, for such applications, smaller carriers will have to be developed for ultrasound-directed drug delivery.

In conclusion, combining ultrasound contrast agents with therapeutic substances may lead to simple and economic methods of treatment with fewer side effects, using conventional ultrasound scanners. Ultrasound-directed drug delivery has great potential in the treatment of malignancies.

---

[55]Fujishiro S, Mitsumori M, Nishimura Y, Okuno Y, Nagata Y, Hiraoka M, Sano T, Marume T, Takayama N. Increased heating efficiency of hyperthermia using an ultrasound contrast agent: a phantom study. *Int J Hyperthermia* **1998** 14:495–502.

[56]Kaneko Y, Maruyama T, Takegami K, Watanabe T, Mitsui H, Hanajiri K, Nagawa H, Matsumoto Y. Use of microbubble agent to increase the effects of high intensity focused ultrasound on liver tissue. *Eur Radiol* **2005** 15:1415–1420.

[57]Tran BC, Seo J, Hall TL, Fowlkes JB, Cain CA. Microbubble-enhanced cavitation for noninvasive ultrasound surgery. *IEEE Trans Ultrason Ferroelectr Freq Control* **2003** 50:1296–1304.

[58]Eisenmenger W. The mechanisms of stone fragmentation in ESWL. *Ultrasound Med Biol* **2001** 27:683–693.

[59]Eisenmenger W, Du XX, Tang C, Zhao S, Wang Y, Rong F, Dai D, Guan M, Qi A. The first clinical results of "wide-focus and low-pressure" ESWL. *Ultrasound Med Biol* **2002** 28:769–774.

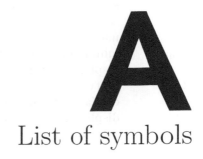

# A
# List of symbols

## Chapter 1

M L T I $\Theta$ dimensions

| | | |
|---|---|---|
| $c_\mathrm{p}$ | phase velocity of a compressional wave | $\mathrm{L\ T^{-1}}$ |
| $d$ | distance between molecules | $\mathrm{L}$ |
| $f$ | frequency | $\mathrm{T^{-1}}$ |
| $k$ | wave number | $\mathrm{L^{-1}}$ |
| | | |
| $\Delta\phi$ | phase difference | |

## Chapter 2

| | | |
|---|---|---|
| $A$ | area | $\mathrm{L^2}$ |
| $A_i$ | area of face $i$ | $\mathrm{L^2}$ |
| $E$ | Young's modulus | $\mathrm{M\ L^{-1}\ T^{-2}}$ |
| $G$ | shear modulus | $\mathrm{M\ L^{-1}\ T^{-2}}$ |
| $I_1$ | first stress invariant | $\mathrm{M\ L^{-1}\ T^{-2}}$ |
| $I_2$ | second stress invariant | $\mathrm{M^2\ L^{-2}\ T^{-4}}$ |
| $I_3$ | third stress invariant | $\mathrm{M^3\ L^{-3}\ T^{-6}}$ |
| $l$ | direction cosine | |
| $l_i$ | direction cosine | |
| $L$ | direction cosine matrix | |
| $m$ | direction cosine | |
| $m_i$ | direction cosine | |

| | | |
|---|---|---|
| $n$ | direction cosine | |
| $n_i$ | direction cosine | |
| $N$ | normal stress | $M\ L^{-1}\ T^{-2}$ |
| $p$ | principal stress | $M\ L^{-1}\ T^{-2}$ |
| $p_i$ | principal stress in direction $i$ | $M\ L^{-1}\ T^{-2}$ |
| $r$ | distance | $L$ |
| $s$ | resultant stress | $M\ L^{-1}\ T^{-2}$ |
| $s_i$ | stress in direction $i$ | $M\ L^{-1}\ T-2$ |
| $T$ | shear stress | $M\ L^{-1}\ T^{-2}$ |
| $T_{\max}$ | maximum shear stress | $M\ L^{-1}\ T^{-2}$ |
| $T_{\min}$ | minimum shear stress | $M\ L^{-1}\ T^{-2}$ |
| $u$ | displacement in the $x$-direction | $L$ |
| $U$ | unity | |
| $v$ | displacement in the $y$-direction | $L$ |
| $w$ | displacement in the $z$-direction | $L$ |
| $x$ | $x$-coordinate | $L$ |
| $y$ | $y$-coordinate | $L$ |
| $z$ | $z$-coordinate | $L$ |
| | | |
| $\gamma_{ij}$ | shear strain on face $i$ in direction $j$ | |
| $\varepsilon$ | strain | |
| $\varepsilon_0$ | octahedral strain | |
| $\varepsilon_i$ | strain in direction $i$ | |
| $\varepsilon_N$ | normal strain | |
| $\varepsilon_p$ | principal strain | |
| $\theta$ | lateral angle | |
| $\kappa$ | bulk modulus | $M\ L^{-1}\ T^{-2}$ |
| $\lambda$ | Lamé's constant | $M\ L^{-1}\ T^{-2}$ |
| $\nu$ | Poisson's ratio | |
| $\sigma$ | stress | $M\ L^{-1}\ T^{-2}$ |
| $\sigma_0$ | octahedral normal stress | $M\ L^{-1}\ T^{-2}$ |
| $\sigma_i$ | stress in direction $i$ | $M\ L^{-1}\ T^{-2}$ |
| $\sigma_{ys}$ | Von Mises yield stress | $M\ L^{-1}\ T^{-2}$ |
| $\tau_0$ | octahedral shear stress | $M\ L^{-1}\ T^{-2}$ |
| $\tau_{ij}$ | shear stress on face $i$ in direction $j$ | $M\ L^{-1}\ T^{-2}$ |
| $\phi$ | stress function | $M\ L\ T^{-2}$ |

# Chapter 3

| | | |
|---|---|---|
| $a$ | acceleration | $L\ T^{-2}$ |
| $A$ | excursion amplitude | $L$ |
| $A_i$ | excursion amplitude | $L$ |
| $b$ | base excursion amplitude | $L$ |
| $B$ | excursion amplitude | $L$ |

| | | |
|---|---|---|
| $c$ | constant | $L^2\,T^{-2}$ |
| c | constant | $T$ |
| $C$ | excursion amplitude | $L$ |
| $E_P$ | potential energy | $M\,L^2\,T^{-2}$ |
| $f(x)$ | variable stiffness function | $M\,L\,T^{-2}$ |
| $f_0$ | resonance frequency | $T^{-1}$ |
| $F$ | force | $M\,L\,T^{-2}$ |
| $F_0$ | force amplitude | $M\,L\,T^{-2}$ |
| $F_i$ | driving function amplitude | $M\,L\,T^{-2}$ |
| $j$ | complex number with the property $j^2 = -1$ | |
| $m$ | mass | $M$ |
| M | magnification factor | |
| $r$ | resistance | $T^{-1}$ |
| $s$ | spring stiffness | $M\,T^{-2}$ |
| $s_1$ | spring stiffness | $M\,T^{-2}$ |
| $s_2$ | spring stiffness | $M\,L^{-2}\,T^{-2}$ |
| $S(\omega, A)$ | response function | $M^2\,L^2\,T^{-4}$ |
| $t$ | time | $T$ |
| $T_0$ | resonance period | $T$ |
| $T_d$ | damped period | $T$ |
| $v$ | velocity | $L\,T^{-1}$ |
| $x$ | $x$-coordinate | $L$ |
| $\dot{x}$ | first time derivative of $x$ | $L\,T^{-1}$ |
| $\ddot{x}$ | second time derivative of $x$ | $L\,T^{-2}$ |
| $\Delta x$ | step size | $L$ |
| $x_0$ | initial position | $L$ |
| $x_b$ | base excursion | $L$ |
| $X$ | steady-state amplitude | $L$ |
| | | |
| $\beta$ | mechanical resistance | $M\,T^{-1}$ |
| $\delta$ | relative excursion amplitude | |
| $\delta_s$ | static deflection | $L$ |
| $\epsilon$ | harmonic displacement | $L^{-2}\,T^{-2}$ |
| $\zeta$ | damping coefficient | |
| $\lambda$ | auxiliary solution | $T^{-2}$ |
| $\phi$ | phase | |
| $\phi$ | phase | |
| $\omega$ | angular frequency | $T^{-1}$ |
| $\omega_0$ | resonance angular frequency | $T^{-1}$ |
| $\omega_d$ | damped natural frequency | $T^{-1}$ |

# Chapter 4

| | | |
|---|---|---|
| $A_i$ | incident pressure amplitude in medium $i$ | $M\,L^{-1}\,T^{-2}$ |

| | | |
|---|---|---|
| $B/A$ | nonlinearity parameter | |
| $B_i$ | reflected pressure amplitude in medium $i$ at a given time | $\mathrm{M\ L^{-1}\ T^{-2}}$ |
| $c$ | sound speed | $\mathrm{L\ T^{-1}}$ |
| $c_i$ | sound speed in medium $i$ | $\mathrm{L\ T^{-1}}$ |
| $c_\mathrm{p}$ | phase velocity of a compressional wave | $\mathrm{L\ T^{-1}}$ |
| $c_\mathrm{s}$ | phase velocity of a transverse wave | $\mathrm{L\ T^{-1}}$ |
| $c_\nu$ | phase velocity | $\mathrm{L\ T^{-1}}$ |
| $d$ | distance | $\mathrm{L}$ |
| $D$ | attenuation | $\mathrm{L^{-1}}$ |
| $E$ | Young's modulus | $\mathrm{M\ L^{-1}\ T^{-2}}$ |
| $E_\mathrm{K}$ | kinetic energy | $\mathrm{M\ L^2\ T^{-2}}$ |
| $E_\mathrm{P}$ | potential energy | $\mathrm{M\ L^2\ T^{-2}}$ |
| $E_\mathrm{T}$ | total energy | $\mathrm{M\ L^2\ T^{-2}}$ |
| $\mathcal{E}$ | energy per unit volume | $\mathrm{M\ L^{-1}\ T^{-2}}$ |
| $f$ | frequency | $\mathrm{T^{-1}}$ |
| $f'$ | experienced frequency | $\mathrm{T^{-1}}$ |
| $f_i(x)$ | pressure function $i$ of $x$ | $\mathrm{M\ L^{-1}\ T^{-2}}$ |
| $f_i''$ | second partial derivative of $f$ | |
| $g_i(x)$ | displacement function $i$ of $x$ | $\mathrm{L}$ |
| $G$ | shear modulus | $\mathrm{M\ L^{-1}\ T^{-2}}$ |
| $i$ | scalar $i \in \mathbb{Z}$ | |
| $I$ | average intensity | $\mathrm{M\ T^{-3}}$ |
| $I_0$ | threshold of hearing intensity | $\mathrm{M\ T^{-3}}$ |
| $I_\mathrm{i}$ | incident intensity | $\mathrm{M\ T^{-3}}$ |
| $I_\mathrm{r}$ | reflected intensity | $\mathrm{M\ T^{-3}}$ |
| $I_t$ | instantaneous intensity | $\mathrm{M\ T^{-3}}$ |
| $I_\mathrm{t}$ | transmitted intensity | $\mathrm{M\ T^{-3}}$ |
| IL | intensity level | |
| $j$ | complex number with the property $j^2 = -1$ | |
| $J_0$ | Bessel function of order zero of the first kind | |
| $k$ | wave number | $\mathrm{L^{-1}}$ |
| $k_i$ | wave number in medium $i$ | $\mathrm{L^{-1}}$ |
| $k_\mathrm{i}$ | wave number of incident wave | $\mathrm{L^{-1}}$ |
| $k_\mathrm{r}$ | wave number of reflected wave | $\mathrm{L^{-1}}$ |
| $k_\mathrm{Re}$ | real wave number | $\mathrm{L^{-1}}$ |
| $k_\mathrm{t}$ | wave number of transmitted wave | $\mathrm{L^{-1}}$ |
| $\mathbf{k}$ | wave vector | $\mathrm{L^{-1}}$ |
| $l$ | interface location | $\mathrm{L}$ |
| $m$ | scalar value | |
| $M$ | molar mass | $\mathrm{M}$ |
| $n$ | amount of gas | |
| n | refraction index | |
| $p$ | acoustic pressure | $\mathrm{M\ L^{-1}\ T^{-2}}$ |
| $p_0$ | pressure amplitude | $\mathrm{M\ L^{-1}\ T^{-2}}$ |
| $p_{0,i}$ | pressure amplitude of wave $i$ | $\mathrm{M\ L^{-1}\ T^{-2}}$ |

| | | |
|---|---|---|
| $p_i$ | acoustic pressure of wave $i$ | M L$^{-1}$ T$^{-2}$ |
| $p_{i,\text{rms}}$ | root-mean-square pressure of wave $i$ | M L$^{-1}$ T$^{-2}$ |
| $p_\text{i}$ | pressure of incident wave | M L$^{-1}$ T$^{-2}$ |
| $p_{f,\text{rms}}$ | root-mean-square pressure at frequency $f$ | M L$^{-1}$ T$^{-2}$ |
| $p_\text{h}$ | hearing threshold | M L$^{-1}$ T$^{-2}$ |
| $p_\text{r}$ | pressure of reflected wave | M L$^{-1}$ T$^{-2}$ |
| $p_\text{rms}$ | root-mean-square pressure | M L$^{-1}$ T$^{-2}$ |
| $p_\text{t}$ | pressure of transmitted wave | M L$^{-1}$ T$^{-2}$ |
| $\mathbf{p}$ | pressure vector | M L$^{-1}$ T$^{-2}$ |
| $\ddot{\mathbf{p}}$ | second time derivative of $\mathbf{p}$ | M L$^{-1}$ T$^{-4}$ |
| $P$ | absolute pressure | M L$^{-1}$ T$^{-2}$ |
| d$P$ | pressure difference | M L$^{-1}$ T$^{-2}$ |
| $P_\text{s}$ | scattered power | M L$^2$ T$^{-3}$ |
| $Q_\text{s}$ | acoustic object size | L$^2$ |
| $r$ | distance | L |
| $r_i$ | distance of location $i$ | L |
| $\mathbf{r}$ | distance vector | L |
| R | reflection coefficient | |
| R$_\text{I}$ | intensity reflection coefficient | |
| $\mathcal{R}$ | gas constant | M L$^2$ T$^{-2}$ $\Theta^{-1}$ |
| $\bar{\mathcal{R}}$ | specific gas constant | L$^2$ T$^{-2}$ $\Theta^{-1}$ |
| $s(t)$ | signal | |
| $S$ | area | L$^2$ |
| $S(\omega)$ | frequency-domain representation | |
| d$S$ | element cross-section | L$^2$ |
| SPL | sound pressure level | |
| SPL$_i$ | sound pressure level at $r_i$ | |
| SWL | sound power level | |
| $t$ | time | T |
| $T$ | period | T |
| T | transmission coefficient | |
| T$_\text{I}$ | intensity transmission coefficient | |
| $\mathcal{T}$ | absolute temperature | $\Theta$ |
| $u$ | displacement in the $x$-direction | L |
| $u_0$ | initial displacement | L |
| $\mathbf{u}$ | displacement vector | L |
| $\ddot{\mathbf{u}}$ | second time derivative of $\mathbf{u}$ | L T$^{-2}$ |
| $U$ | unity | |
| $V$ | volume | L$^3$ |
| d$V$ | volume difference | L$^3$ |
| $V_0$ | initial volume | L$^3$ |
| $W$ | sound power | M L$^2$ T$^{-3}$ |
| $W_0$ | threshold of hearing power | M L$^2$ T$^{-3}$ |
| $x$ | $x$-coordinate | L |
| d$x$ | element length | L |
| $x_0$ | initial position | L |

| | | |
|---|---|---|
| $x_i$ | position at time $i$ | L |
| $x_\infty$ | saw-tooth distance | L |
| $Z$ | acoustic impedance | M L$^{-2}$ T$^{-1}$ |
| $Z_\mathrm{c}$ | relative characteristic impedance | |
| $Z_i$ | acoustic impedance of medium $i$ | M L$^{-2}$ T$^{-1}$ |
| $Z_\mathrm{s}$ | relative surface impedance | |
| | | |
| $\alpha$ | attenuation coefficient | L$^{-1}$ |
| $\alpha_\mathrm{v}$ | viscous damping coefficient | L$^{-1}$ |
| $\alpha_\theta$ | thermal damping coefficient | L$^{-1}$ |
| $\beta_\mathrm{c}$ | relative characteristic admittance | |
| $\gamma$ | ratio of specific heats | |
| $\gamma_{ij}$ | shear strain on face $i$ in direction $j$ | |
| $\Delta$ | volumetric strain | |
| $\varepsilon$ | strain | |
| $\varepsilon_i$ | strain in direction $i$ | |
| $\eta$ | dynamic viscosity | M L$^{-1}$ T$^{-1}$ |
| $\eta(\omega)$ | backscattering coefficient | |
| $\theta$ | aperture | |
| $\theta_i$ | angle of incidence in medium $i$ | |
| $\theta_\mathrm{i}$ | angle of incidence | |
| $\theta_\mathrm{r}$ | angle of reflection | |
| $\theta_\mathrm{t}$ | angle of transmission | |
| $\kappa$ | bulk modulus | M L$^{-1}$ T$^{-2}$ |
| $\kappa_i$ | compressibility of medium $i$ | M L$^{-1}$ T$^{-2}$ |
| $\lambda$ | Lamé's constant | M L$^{-1}$ T$^{-2}$ |
| $\lambda$ | wavelength | L |
| $\lambda'$ | reduced wavelength | L |
| $\lambda_i$ | wavelength in medium $i$ | L |
| $\nu$ | particle velocity | L T$^{-1}$ |
| $\nu_0$ | particle velocity amplitude | L T$^{-1}$ |
| $\nu_\mathrm{a}$ | velocity of the audience | L T$^{-1}$ |
| $\nu_\mathrm{i}$ | phase velocity of incident wave | L T$^{-1}$ |
| $\nu_\mathrm{r}$ | phase velocity of reflected wave | L T$^{-1}$ |
| $\nu_\mathrm{s}$ | velocity of sound source | L T$^{-1}$ |
| $\nu_\mathrm{t}$ | phase velocity of transmitted wave | L T$^{-1}$ |
| $\Pi$ | disjoining pressure | M L$^{-1}$ T$^{-2}$ |
| $\rho$ | density | M L$^{-3}$ |
| $\rho_i$ | density of medium $i$ | M L$^{-3}$ |
| $\sigma$ | stress | M L$^{-1}$ T$^{-2}$ |
| $\sigma_i$ | stress in direction $i$ | M L$^{-1}$ T$^{-2}$ |
| $\tau_{ij}$ | shear stress on face $i$ in direction $j$ | M L$^{-1}$ T$^{-2}$ |
| $\phi$ | scalar potential | L$^2$ |
| $\phi_i$ | phase of wave $i$ | |
| $\psi$ | vector potential | L$^2$ |
| $\omega$ | angular frequency | T$^{-1}$ |

| $\omega_i$ | angular frequency $i$ | $T^{-1}$ |
| $\Omega$ | beats frequency | $T^{-1}$ |

# Chapter 5

| $a$ | spacing | $L$ |
| $A_i$ | incident pressure amplitude in medium $i$ | $M\,L^{-1}\,T^{-2}$ |
| $b$ | spacing | $L$ |
| $B_i$ | reflected pressure amplitude in medium $i$ at a given time | $M\,L^{-1}\,T^{-2}$ |
| $BW$ | bandwidth | $T^{-1}$ |
| $c$ | elastic stiffness constant | $M\,L^{-1}\,T^{-2}$ |
| $C$ | capacitance | $M^{-1}\,L^{-2}\,T^4\,I^2$ |
| $d$ | piezo-electric strain/charge constant | $M^{-1}\,L^{-1}\,T^3\,I$ |
| $D$ | electric displacement | $L^{-2}\,T\,I$ |
| D | electric displacement (superscript) | |
| $e$ | piezo-electric stress constant | $L^{-2}\,T\,I$ |
| $E$ | electric field | $M\,L\,T^{-3}I^{-1}$ |
| E | electric field (superscript) | |
| $f_{\mathrm{c}}$ | centre frequency | $T^{-1}$ |
| $f_{\mathrm{l}}$ | lower cut-off frequency | $T^{-1}$ |
| $f_{\mathrm{u}}$ | upper cut-off frequency | $T^{-1}$ |
| $F$ | force | $M\,L\,T^{-2}$ |
| $FBW$ | fractional bandwidth | |
| $g$ | piezo-electric voltage constant | $L^2\,T^{-1}\,I^{-1}$ |
| HIFU | high-intensity focussed ultrasound | |
| $I$ | current | $I$ |
| $j$ | complex number with the property $j^2 = -1$ | |
| $k$ | electro-mechanical coupling constant | |
| $k_i$ | wave number of medium $i$ | $L^{-1}$ |
| $l$ | thickness | $L$ |
| $L$ | inductance | $M\,L^2\,T^{-2}\,I^{-2}$ |
| LE | length expander | |
| $\mathcal{L}$ | spacing | $L$ |
| $n$ | turns ratio | |
| $N$ | frequency constants | $L\,T^{-1}$ |
| NDT | non-destructive testing | |
| $P$ | polarisation | $L^{-2}\,T\,I$ |
| PMN | lead metaniobate | |
| PT | lead titanate | |
| PVDF | polymer polyvinylidene fluoride | |
| PZT | lead zirconate titanate | |
| $q$ | charge | $T\,I$ |

| | | |
|---|---|---|
| $Q_\mathrm{m}$ | mechanical Q-factor | |
| $R$ | resistance | $\mathrm{M\ L^2\ T^{-3}\ I^{-2}}$ |
| R | reflection coefficient | |
| RE | radial expander | |
| $s$ | elastic compliance constant | $\mathrm{M^{-1}\ L\ T^2}$ |
| $S$ | strain | |
| S | strain (superscript) | |
| $T$ | stress | $\mathrm{M\ L^{-1}\ T^{-2}}$ |
| T | stress (superscript) | |
| TE | thickness expander | |
| $V$ | voltage | $\mathrm{M\ L^2\ T^{-3}\ I^{-1}}$ |
| WE | width expander | |
| $Y$ | Young's modulus | $\mathrm{M\ L^{-1}\ T^{-2}}$ |
| $Y_i$ | admittance of medium $i$ | $\mathrm{M^{-1}\ L^{-2}\ T^3\ I^2}$ |
| $Z_i$ | acoustic impedance of medium $i$ | $\mathrm{M\ L^{-2}\ T^{-1}}$ |
| $Z_\mathrm{i}$ | electrical impedance of component i | $\mathrm{M\ L^2\ T^{-3}\ I^{-2}}$ |
| | | |
| $\epsilon$ | dielectric constant | $\mathrm{M^{-1}\ L^{-3}\ T^4\ I^2}$ |
| $\lambda$ | wavelength | $\mathrm{L}$ |
| $\rho$ | density | $\mathrm{M\ L^{-3}}$ |
| $\phi$ | phase | |
| $\omega$ | angular frequency | $\mathrm{T^{-1}}$ |

# Chapter 6

| | | |
|---|---|---|
| $a$ | piston radius | $\mathrm{L}$ |
| $A$ | radius of curvature | $\mathrm{L}$ |
| $\Delta A$ | elemental size | $\mathrm{L^2}$ |
| $AW$ | aperture width | $\mathrm{L}$ |
| $b$ | $b = A - \sqrt{A^2 - a^2}$ | $\mathrm{L}$ |
| $B$ | $B(z) = \sqrt{z^2 + 2b(A - z)}$ | $\mathrm{L}$ |
| $c$ | speed of sound | $\mathrm{L\ T^{-1}}$ |
| CW | continuous wave | |
| $d$ | inter-element spacing | $\mathrm{L}$ |
| $D$ | directivity of an array | $\mathrm{L^{-2}}$ |
| $F\#$ | $F$-number | |
| $F_\mathrm{geo}$ | geometric focal length | $\mathrm{L}$ |
| $Fn$ | $F$-number | |
| $g$ | Green's function | $\mathrm{L^{-1}\ T^{-1}}$ |
| $G_\mathrm{p}$ | pressure focal gain | |
| $h$ | impulse response function | $\mathrm{L\ T^{-1}}$ |
| $\Delta h$ | height | $\mathrm{L}$ |
| HIFU | high-intensity focussed ultrasound | |
| $j$ | complex number with the property $j^2 = -1$ | |

| $J_i$ | Bessel function of order $i$ of the first kind | |
| $k$ | wave number | $\text{L}^{-1}$ |
| $L$ | length | L |
| $L_i$ | length in direction $i$ | L |
| $n$ | scalar $n \in \mathbb{N}$ | |
| $N$ | number of elements | |
| $p$ | pressure | $\text{M L}^{-1}\,\text{T}^{-2}$ |
| $P_0$ | pressure amplitude | $\text{M L}^{-1}\,\text{T}^{-2}$ |
| $r$ | distance | L |
| $S$ | surface | $\text{L}^2$ |
| $t$ | time | T |
| $\Delta t$ | time delay of element $i$ | T |
| $v_0$ | particle velocity | $\text{L T}^{-1}$ |
| $\Delta w$ | width | L |
| $x$ | $x$-coordinate | L |
| $y$ | $y$-coordinate | L |
| $z$ | $z$-coordinate | L |
| $z_\text{f}$ | focal depth | L |
| $Z_0$ | acoustic impedance | $\text{M L}^{-2}\,\text{T}^{-1}$ |
| | | |
| $\gamma$ | $\gamma$-coordinate | |
| $\delta$ | Kronecker delta function | $\text{T}^{-1}$ |
| $\eta$ | $\eta$-coordinate | |
| $\theta$ | $\theta$-coordinate | |
| $\rho$ | $\rho$-coordinate | L |
| $\rho_0$ | density | $\text{M L}^{-3}$ |
| $\phi$ | velocity potential | $\text{L}^2\,\text{T}^{-1}$ |
| $\omega$ | angular frequency | $\text{T}^{-1}$ |
| $\Omega$ | $\Omega$-function | |

# Chapter 7

| $A$ | area | $\text{L}^2$ |
| A | amplitude | |
| ALARA | As Low As Reasonably Achievable | |
| $B$ | pulse bandwidth | $\text{T}^{-1}$ |
| B | brightness | |
| $c$ | sound speed | $\text{L T}^{-1}$ |
| CD | colour Doppler | |
| CFM | colour flow mapping | |
| CWD | continuous wave Doppler | |
| $d$ | depth | L |
| $D$ | aperture | L |
| $f_0$ | transmit frequency | $\text{T}^{-1}$ |

| $f_c$ | centre frequency | $T^{-1}$ |
|---|---|---|
| $f_D$ | Doppler shift | $T^{-1}$ |
| $f_{D,max}$ | maximum Doppler shift | $T^{-1}$ |
| $F$ | focal depth | $L$ |
| HPRF | high pulse repetition frequency | |
| $\mathcal{L}$ | lateral resolution | $L$ |
| LPRF | low pulse repetition frequency | |
| M | motion | |
| MI | mechanical index | |
| MRI | magnetic resonance imaging | |
| $p^-$ | peak negative pressure | $M\,L^{-1}\,T^{-2}$ |
| PRF | pulse repetition frequency | $T^{-1}$ |
| PWD | pulsed wave Doppler | |
| $Q$ | flow | $L^3\,T^{-1}$ |
| $R_{max}$ | maximum distance | $L$ |
| RF | radio frequency | |
| $\mathcal{R}$ | resolution | $L$ |
| SATA | spatial average/temporal average | |
| SPTA | spatial peak/temporal average | |
| $t$ | time | $T$ |
| TGC | time gain compensation | |
| THI | tissue harmonic imaging | |
| TI | thermal index | |
| TIB | thermal index of bone | |
| TIC | thermal index of cranial bone | |
| TIS | thermal index of soft tissue | |
| TVG | time varying gain | |
| $v$ | velocity | $L\,T^{-1}$ |
| $v_{max}$ | maximum velocity | $L\,T^{-1}$ |
| $W$ | power | $M\,L^2\,T^{-3}$ |
| $W_{deg}$ | power needed to raise the temperature by 1°C | $M\,L^2\,T^{-3}$ |
| $z$ | $z$-coordinate | $L$ |
| | | |
| $\theta$ | angle | |
| $\lambda$ | wavelength | $L$ |
| $\tau$ | pulse length | $T$ |

# Chapter 8

| $A$ | area | $L^2$ |
|---|---|---|
| $C$ | mass concentration | $M\,L^{-3}$ |
| $C_0$ | saturation mass concentration | $M\,L^{-3}$ |
| $C_i$ | initial mass concentration | $M\,L^{-3}$ |
| $C_s$ | surface saturation mass concentration | $M\,L^{-3}$ |

| | | |
|---|---|---|
| $D$ | dissolution constant | $L^2\ T^{-1}$ |
| $E_{f,i}$ | fragment surface free energy | $M\ L^2\ T^{-2}$ |
| $E_k$ | kinetic energy | $M\ L^2\ T^{-2}$ |
| $E_s$ | surface free energy | $M\ L^2\ T^{-2}$ |
| f | equilibrium density ratio | |
| $F$ | force | $M\ L\ T^{-2}$ |
| $h$ | thickness | $L$ |
| $h_c$ | critical thickness | $L$ |
| $h_i$ | initial thickness | $L$ |
| $k$ | wave number | $L$ |
| $k_g$ | Henry's constant | $L^2\ T^{-2}$ |
| k | Boltzmann's constant | $M\ L^2\ T^{-2}\ \Theta^{-1}$ |
| $l_j$ | jet length | $L$ |
| $L$ | Ostwald's solubility coefficient | |
| $m$ | mass | $M$ |
| MI | mechanical index | |
| $n$ | oscillation mode | |
| $N$ | number of fragments | |
| $p$ | pressure | $M\ L^{-1}\ T^{-2}$ |
| $\Delta p$ | overpressure | $M\ L^{-1}\ T^{-2}$ |
| $p_0$ | ambient pressure | $M\ L^{-1}\ T^{-2}$ |
| $p_0^\infty$ | pressure at infinity | $M\ L^{-1}\ T^{-2}$ |
| $p_{cr}$ | critical pressure | $M\ L^{-1}\ T^{-2}$ |
| $p_g$ | gas pressure | $M\ L^{-1}\ T^{-2}$ |
| $p_i$ | pressure inside bubble $i$ | $M\ L^{-1}\ T^{-2}$ |
| $p_L$ | liquid pressure | $M\ L^{-1}\ T^{-2}$ |
| $p_{LY}$ | Laplace–Young pressure | $M\ L^{-1}\ T^{-2}$ |
| $p_n$ | pressure in situation $n$ | $M\ L^{-1}\ T^{-2}$ |
| $p_v$ | vapour pressure | $M\ L^{-1}\ T^{-2}$ |
| $p_{wh}$ | water hammer pressure | $M\ L^{-1}\ T^{-2}$ |
| $P(t)$ | driving function | $M\ L^{-1}\ T^{-2}$ |
| $P_A$ | pressure amplitude | $M\ L^{-1}\ T^{-2}$ |
| $r$ | distance | $L$ |
| $\dot{r}$ | particle velocity | $L\ T^{-1}$ |
| $R$ | radius | $L$ |
| $\dot{R}$ | surface velocity | $L\ T^{-1}$ |
| $\ddot{R}$ | surface acceleration | $L\ T^{-2}$ |
| $R_0$ | quasi-equilibrium radius | $L$ |
| $R_c$ | collapse radius | $L$ |
| $R_{cr}$ | critical radius | $L$ |
| $R_f$ | film radius | $L$ |
| $R_{f,m}$ | mean fragment radius | $L$ |
| $R_i$ | radius of bubble $i$ | $L$ |
| $\dot{R}_i$ | surface velocity of bubble $i$ | $L\ T^{-1}$ |
| $R_i$ | inner radius | $L$ |

| $\dot{R}_\mathrm{i}$ | inner surface velocity | $\mathrm{L\ T^{-1}}$ |
|---|---|---|
| $R_\mathrm{m}$ | mean radius | $\mathrm{L}$ |
| $\mathcal{R}$ | gas constant | $\mathrm{M\ L^2\ T^{-2}\ \Theta^{-1}}$ |
| $\bar{\mathcal{R}}$ | specific gas constant | $\mathrm{L^2\ T^{-2}\ \Theta^{-1}}$ |
| $S$ | path | $\mathrm{L}$ |
| $t$ | time | $\mathrm{T}$ |
| $\Delta t$ | time step | $\mathrm{T}$ |
| $T$ | period | $\mathrm{T}$ |
| $\mathcal{T}$ | absolute temperature | $\Theta$ |
| $u$ | velocity | $\mathrm{L\ T^{-1}}$ |
| $\dot{u}$ | acceleration | $\mathrm{L\ T^{-2}}$ |
| $v$ | fluid velocity | $\mathrm{L\ T^{-1}}$ |
| $\dot{v}$ | fluid acceleration | $\mathrm{L\ T^{-2}}$ |
| $v_\mathrm{j}$ | jet velocity | $\mathrm{L\ T^{-1}}$ |
| $V$ | volume | $\mathrm{L^3}$ |
| $\Delta V$ | expansion amplitude | $\mathrm{L^3}$ |
| $V_0$ | quasi-equilibrium volume | $\mathrm{L^3}$ |
| $V_i$ | equilibrium volume of bubble $i$ | $\mathrm{L^3}$ |
| $\Delta V_i$ | expansion amplitude of bubble $i$ | $\mathrm{L^3}$ |
| $V_\mathrm{j}$ | jet volume | $\mathrm{L^3}$ |
| $V_n$ | volume in situation $n$ | $\mathrm{L^3}$ |
| We | Weber number | |
| $x$ | $x$-coordinate | $\mathrm{L}$ |
| | | |
| $\gamma$ | ratio of specific heats | |
| $\dot{\varepsilon}$ | rate of strain | $\mathrm{T^{-1}}$ |
| $\dot{\varepsilon}_\mathrm{r}$ | radial rate of strain | $\mathrm{T^{-1}}$ |
| $\zeta$ | damping coefficient | |
| $\zeta_\mathrm{s}$ | shell damping coefficient | |
| $\zeta_\mathrm{v}$ | viscous damping coefficient | |
| $\eta$ | dynamic viscosity | $\mathrm{M\ L^{-1}\ T^{-1}}$ |
| $\xi$ | oscillation amplitude | $\mathrm{L}$ |
| $\rho$ | density | $\mathrm{M\ L^{-3}}$ |
| $\rho_\mathrm{g}$ | gas density | $\mathrm{M\ L^{-3}}$ |
| $\rho_i$ | equilibrium density of bubble $i$ | $\mathrm{M\ L^{-3}}$ |
| $\rho_\mathrm{s}$ | shell density | $\mathrm{M\ L^{-3}}$ |
| $\sigma$ | surface tension | $\mathrm{M\ L^{-1}\ T^{-2}}$ |
| $\sigma_i$ | surface tension at interface $i$ | $\mathrm{M\ L^{-1}\ T^{-2}}$ |
| $\tau_\mathrm{d}$ | drainage time | $\mathrm{T}$ |
| $\phi$ | phase difference | |
| $\Phi$ | velocity potential | $\mathrm{L^2\ T^{-1}}$ |
| $\omega$ | angular frequency | $\mathrm{T^{-1}}$ |
| $\omega_0$ | angular resonance frequency | $\mathrm{T^{-1}}$ |

# Chapter 9

| CEUS | contrast-enhanced ultrasound |
|------|------------------------------|
| DNA  | deoxyribonucleic acid |
| ESWL | extracorporeal shock wave lithotripsy |
| GI   | gastrointestinal |
| HIFU | high-intensity focussed ultrasound |

# Symbols

$\nabla \phi$      $\operatorname{grad} \phi = \left( \frac{\partial \phi}{\partial x} \ \ \frac{\partial \phi}{\partial y} \ \ \frac{\partial \phi}{\partial z} \right)^{\mathrm{T}}$

$\nabla^2 \phi$      $\nabla \cdot \nabla \phi = \operatorname{div} \operatorname{grad} \phi$

$\qquad\qquad = \frac{\partial^2 \phi}{\partial x^2} + \frac{\partial^2 \phi}{\partial y^2} + \frac{\partial^2 \phi}{\partial z^2}$ (Laplacian)

$\nabla \cdot \boldsymbol{\phi}$      $\operatorname{div} \boldsymbol{\phi}$

$\nabla \times \boldsymbol{\phi}$      $\operatorname{curl} \boldsymbol{\phi}$

# B

# Recommended reading

1. Alonso M, Finn EJ. *Physics*. Harlow: Pearson **1992**.

2. Atchley AA, Sparrow VW, Keolian RM, Eds. *Innovations in Nonlinear Acoustics*. Melville: American Institute of Physics **2006**.

3. Beatty MF. *Principles of Engineering Mechanics: Volume 2. Dynamics — The Analysis of Motion*. New York: Springer **2006**.

4. Bernoulli D. *Hydrodynamica, sive de viribus et motibus fluidorum commentarii*. Strasbourg: JH Dulsecker **1738**.

5. Brennen CE. *Cavitation and bubble dynamics*. Oxford: Oxford University Press **1995**.

6. Breyer RT. *Nonlinear Acoustics*. Woodbury: Acoustical Society of America **1997**.

7. Dury G, Jones D, Thorp S, Eds. *Roger's Profanisaurus*. London: Dennis **2007**.

8. Fletcher NH, Rossing TD. *The Physics of Musical Instruments*. 2nd ed. New York: Springer **1998**.

9. Gere JM. *Mechanics of Materials*. 6th ed. Toronto: Thomson **2006**.

10. Hibbeler RC. *Engineering Mechanics DYNAMICS*. 11th ed. in SI units. Singapore: Pearson Prentice Hall **2007**.

11. Hill CR, Bamber JC, ter Haar GR, Eds. *Physical Principles of Medical Ultrasonics*. 2nd ed. Hoboken: Wiley **2004**.

234

12. Horowitz P, Hill W. *The art of electronics.* Cambridge: Cambridge University Press **1980**.

13. Kuttruff H. *Acoustics: An introduction.* London: Taylor & Francis **2007**.

14. Landau LD, Lifshitz EM. *Fluid Mechanics.* 2nd ed. Oxford: Butterworth-Heinemann **1987**.

15. Landau LD, Lifshitz EM. *Theory of Elasticity.* 3rd ed. Oxford: Butterworth-Heinemann **1986**.

16. Lauterborn W, Kurz T, Eds. *Nonlinear Acoustics at the Turn of the Millennium.* New York: American Institute of Physics **2000**.

17. Leighton TG. *The Acoustic Bubble.* London: Academic Press **1994**.

18. Meriam JL, Kraige LG. *Engineering Mechanics: Volume 2. Dynamics.* 3rd ed. Chichester: Wiley **1993**.

19. Pain HJ. *The Physics of Vibrations and Waves.* 3rd ed. Chichester: Wiley **1983**.

20. Pierce AD. *Acoustics: An Introduction to Its Physical Principles and Applications.* 1994 ed. Melville: Acoustical Society of America **1994**.

21. Porges G. *Applied Acoustics.* Los Altos: Peninsula **1977**.

22. Lord Rayleigh. *The Theory of Sound.* Vol. 1, 2nd ed. Dover **1945**.

23. Lord Rayleigh. *The Theory of Sound.* Vol. 2, 2nd ed. Dover **1945**.

24. Stöcker H, Ed. *Taschenbuch der Physik.* 4th ed. Thun: Deutsch **2000**.

25. Webb A. *Introduction to Biomedical Imaging.* Hoboken: Wiley **2003**.

26. Wright MCM, Ed. *Lecture Notes on the Mathematics of Acoustics.* London: Imperial College Press **2005**.

27. Ødegaard S, Nesje LB, Gilja OH. *Atlas of endoscopic ultrasonography: the upper gastrointestinal tract.* Bergen: Fagbokforlaget **2007**.

# Biographies

Prof. Dr. Michiel Postema, FIOA, was born in Brederwiede, Netherlands, in 1973. He received an M.Sc. in Geophysics from Utrecht University, Netherlands, in 1996 and a Doctorate in the Physics of Fluids from the University of Twente, Enschede, Netherlands, in 2004. Following a postdoctoral position at Ruhr-Universität Bochum, Germany, between 2005 and 2007, he became Lecturer in Engineering at The University of Hull, England. He was granted an Emmy Noether Research Group at Ruhr-Universität Bochum in 2009 and a Visiting Professorship at L'Université d'Orléans, France, in 2010. In the same year, he obtained the Chair in Experimental Acoustics at the University of Bergen, Norway. He has written 60 scientific publications, including 40 first-author papers and five co-authored textbooks. His particular expertise lies in analysing medical microbubble behaviour under sonication and in high-speed photography. He also explores non-medical applications of bubbles and droplets in sound fields.

Prof. Dr. Keith Attenborough, CEng, FIOA, FASA, was born in Nottingham, England, in 1944. He received a B.Sc. in Physics from University College London in 1965 and a Doctorate in Applied Science from the University of Leeds, England, in 1969. Following a postdoctoral position at the University of Liverpool between 1969 and 1970, he became Lecturer in Engineering Mechanics at The Open University, United Kingdom. He was granted a Personal Chair in Acoustics in 1992 and received the Rayleigh Medal of the Institute of Acoustics in 1996. In 1998 he became Professor and Head of the School of Engineering at The University of Hull. Following early retirement from Hull, where he is now Emeritus Professor, he returned to the Open University as a part-time Research Professor. He has written more than 100 scientific publications, including three co-authored textbooks. His particular expertise lies in aspects of outdoor sound propagation but he has worked also on theories of sound propagation in porous materials, including cancellous bone. His current research projects include sonic crystals, acoustical sensing of soils and exploitation of natural means for noise control.

Prof. Dr. Michael J. Fagan was born on the Isle of Wight, England, in 1957. He received a first class B.Sc. in Engineering Science from the University of Exeter in 1979. After working for a few years in British Aerospace and Westland Helicopters, he returned to the University of Exeter to study for his Doctorate in Medical Engineering, researching into the design of artificial hip joints. He then joined The University of Hull as a Lecturer and was granted a Chair in Medical and Biological Engineering in 2008. He now leads the Medical and Biological Engineering Research Group, which focuses on modelling and simulation in biomechanics and biomedical engineering, with a particular interest in understanding the mechanobiology of bone, using finite element analysis and multibody dynamics analysis.

Prof. Dr. Odd Helge Gilja was born in Bergen, Norway, in 1962. He received a Medical Degree from the University of Bergen in 1990 and a Ph.D. from the University of Bergen in 1997. He specialised in internal medicine and gastroenterology at Haukeland University Hospital, Bergen. He was appointed Associate Professor at the Institute of Medicine, University of Bergen, in 2001, and Professor in 2002. He was president of the Norwegian Society for Diagnostic Ultrasound in Medicine from 2001 to 2007. He has been Advisory Board member of European Federation for Ultrasound in Medicine and Biology (EFSUMB) since 2001. He has been appointed Chairman of EFSUMB's Education and Professional Standards Committee in 2007. He has written more than 100 scientific publications, including 25 book chapters and four edited books. He has twice won the EFSUMB European Young Investigators Award. In 2006, he was elected to give the prestigious Euroson Lecture in Bologna. His present positions are: Professor at the Institute of Medicine, University of Bergen; Senior Consultant at the National Centre for Ultrasound in Gastroenterology, Haukeland University Hospital; Director of MedViz — an interdisciplinary Research & Development cluster in medical imaging and visualisation. His particular expertise is abdominal ultrasound scanning, 3D ultrasound, strain imaging, contrast-enhanced ultrasound, gastric motility, and ultrasound visualisation.

Dr. Andrew Hurrell was born in Ashford, Kent, England, in 1972. He received a B.Sc. (Hons) in Physics and Modern Acoustics from Surrey University, England, in 1994 and the Institute of Acoustics (IOA) Diploma in Acoustics and Noise Control Engineering in the same year. Upon graduation in 1994 he joined the Defence Research Agency in the Acoustics Materials Section. In 1996 he joined Precision Acoustics Ltd, Dorchester, England, as a senior research physicist — a position that he still holds. Whilst working for Precision Acoustics, he also was registered as a part time student through the University of Bath, and in 2002 was awarded a PhD in Underwater Acoustics. Since 2002, he has represented UK national interest on the IEC Technical Committee 87 (Ultrasonics), and has contributed to the development of numerous international standards in acoustic and ultrasonic metrology. He has written more than 35 scientific publications. In 2005 he won the IOA Young Persons' Award for Innovation in Acoustical Engineering. His scientific expertise lies in the development of novel ultrasonic sensors and transducers for both medical and industrial applications, and the numerical modelling of ultrasonic propagation.

Mr. Spiros Kotopoulis was born in Athens, Greece, in 1987. He received a B.Eng. (Hons) in Mechanical Engineering from The University of Hull, Kingston upon Hull, England, in 2008. He is currently pursuing a Ph.D. in the Department of Engineering at the University of Hull under the supervision of Prof. Dr. Michiel Postema. His particular expertise lies in high-speed photography, microscopy, and ultrasound transducer manufacture.

Prof. Dr. Knut Matre was born in Lillehammer, Norway, in 1950. He received a B.Sc. in Applied Physics from Heriot-Watt University, Edinburgh, Scotland, in 1976 and an M.Sc. in Medical Physics from the University of Aberdeen, Scotland, in 1977. He spent two years working on ultrasound Doppler methods at the Norwegian University of Science and Technology in Trondheim and came to the University of Bergen in 1979. He received a Dr.philos. degree on invasive ultrasound from the Surgical Research Laboratory, University of Bergen, in 1988. After being an Assistant Professor at the Institute of Medicine, University of Bergen, between 1986 and 2002, he became Professor in Medical Physics and Technology. His research interests include methology and analysis of myocardial ischemia, development and evaluation of new cardiac ultrasound methods, studies of the velocity distribution in heart chambers and large arteries, and development of new ultrasound methods in gastroenterology and nephrology.

Dr. Annemieke (J.E.T.) van Wamel was born in Groesbeek, Netherlands, in 1969. She received an M.Sc. in Biology (Biochemistry and Molecular Plant Physiology) from Radboud University Nijmegen, Netherlands, in 1994. She obtained a Ph.D. degree on intracardiac tissue remodelling as a response to myocardial infarction at the Cardiology Department of the Leiden University Medical Centre, Netherlands, in 2001. As leading scientist she worked on molecular imaging of (cardio)vascular disease using ultrasound contrast agents, ultrasound-controlled local drug and cell delivery, and ultrasound-induced cellular/tissue responses at the Department of Biomedical Engineering at Erasmus MC Rotterdam, Netherlands, from 2001 until 2009. She was visiting researcher in the laboratory of Dr. Alexander L. Klibanov at the University of Virginia, USA, from 2007 to 2008. From 2008 to 2009, she was a member of the European Committee for Medical Ultrasound Safety (ECMUS). She has written more than 100 scientific publications. Currently, she teaches biology and life science at college preparatory high schools in the Netherlands.

Dr. Paul A. Campbell, FInstP, FRAS, was born in Belfast, Northern Ireland, in 1968. He read Physics as an undergraduate at the University of London — Queen Mary College (1990), before taking a Ph.D. in Experimental Physics at Queen's University Belfast (1994). After a brief period in the defence industry, he subsequently gained invaluable postdoctoral experience in both hospital-, and university-based Applied Physics groups. He became active in ultrasound-related research during 2003, having been influenced hugely during a period as a visiting scholar working with Professor Mark Prausnitz at the Georgia Institute of Technology in Atlanta. His present research is focussed on understanding the fundamental microscopic interactions of ultrasound waves with biological cells and tissues, particularly in the presence of microbubbles. He is presently a Reader in Physics at the Carnegie Physics Laboratory at the University of Dundee, Scotland, and also deputy head of the Division of Molecular Medicine; he holds active research funding in the capacity of PI at the level of over £2.2M; has written some 70 journal and conference papers; and, from October 2010, will be a Royal Society Industry Fellow working closely with a small spin-out company based at Cambridge University. His strategic research programme has the ultimate objective of achieving reliable and non-invasive drug delivery in clinical settings.

# Index